T0336182

AUTONOMY AND TRUST IN BIOETHICS

Why has autonomy been a leading idea in philosophical writing on bioethics, and why has trust been marginal? In this important book, Onora O'Neill suggests that the conceptions of individual autonomy so widely relied on in bioethics are philosophically and ethically inadequate, and that they undermine, rather than support, relations of trust. She shows how Kant's non-individualistic view of autonomy provides a stronger basis for an approach to medicine, science and biotechnology, and does not marginalise trust, while also explaining why trustworthy individuals and institutions are often undeservingly mistrusted. Her arguments are illustrated with issues raised by practices such as the use of genetic information by the police or insurers, research using human tissues, uses of new reproductive technologies and media practices for reporting on medicine, science and technology. *Autonomy and Trust in Bioethics* will appeal to a wide range of readers in ethics, bioethics and related disciplines.

ONORA O'NEILL is Principal of Newnham College, Cambridge. She has written widely on ethics and political philosophy, with particular focus on questions of international justice, the philosophy of Kant and bioethics. Her most recent books include *Towards Justice and Virtue: A Constructive Account of Practical Reasoning* (Cambridge, 1996) and *Bounds of Justice* (Cambridge, 2000).

AUTONOMY AND TRUST IN BIOETHICS

The Gifford Lectures
University of Edinburgh, 2001

ONORA O'NEILL
Newnham College, Cambridge

CAMBRIDGE UNIVERSITY PRESS
Cambridge, New York, Melbourne, Madrid, Cape Town, Singapore, São Paulo

Cambridge University Press
The Edinburgh Building, Cambridge CB2 8RU, UK

Published in the United States of America by Cambridge University Press, New York

www.cambridge.org
Information on this title: www.cambridge.org/9780521815406

First published 2002
Third printing 2005

A catalogue record for this publication is available from the British Library

ISBN 978-0-521-81540-6 hardback
ISBN 978-0-521-89453-1 paperback

Transferred to digital printing 2007

Contents

Preface

Autonomy has been a leading idea in philosophical writing on bioethics; trust has been marginal. This strikes me as surprising. Autonomy is usually identified with individual independence, and sometimes leads to ethically dubious or disastrous action. Its ethical credentials are not self-evident. Trust is surely more important, and particularly so for any ethically adequate practice of medicine, science and biotechnology. Trust – or rather loss of trust – is a constant concern in political and popular writing in all three areas. Why then has autonomy landed a starring role in philosophical and ethical writing in bioethics? And why has trust secured no more than a walk-on part?

When I was invited to deliver the Gifford Lectures for 2001 in the University of Edinburgh, I rashly chose the title *Autonomy and Trust in Bioethics*. I was interested in this divergence between philosophical and popular ethical concerns, and the reasons for its persistence. The topic proved fruitful and more recalcitrant than I had expected. With the help of a thoughtful and encouraging audience in Edinburgh, and of numerous suggestions and comments from friends and colleagues, I have explored a wider terrain than I had originally intended. I have come to think that many recent discussions of both autonomy and of trust are unconvincing, and that this matters greatly for the ways in which we think about ethical questions that arise in the practice of medicine, science and biotechnology. Discussions of autonomy and trust in other areas of life may also be unconvincing; but that is a topic for another occasion.

Although I have been critical of contemporary work in bioethics in this book, my aims are both philosophically and practically constructive. They are philosophically constructive in that I set out and state the case for a conception of practical reasoning that supports a wide range of robust ethical obligations, ranging from requirements to seek informed consent (devotees of individual autonomy have been right to stress them) to practices that secure trustworthiness and may support relations of trust. They are practically constructive in that I show how these requirements are relevant to many areas of controversy, extending from public policy to the regulation of medicine, science and reproductive technologies, to daily medical and scientific practice and the uses of biotechnologies.

Writing on bioethics exacts intellectually troubling compromises. If it is to be philosophically serious it cannot take specific institutional and professional arrangements for granted; if it is to speak to actual predicaments it must take institutional and professional arrangements seriously. Much writing on bioethics fails as philosophy because it takes for granted some of the institutions or practices of particular cultures or times, such as hospital-based medicine or advanced biotechnologies, and fails to consider alternatives. Some philosophically interesting writing lacks clear implications for medicine, science and biotechnology because it is oblivious to institutional and professional realities and diversities. These problems can be avoided but not solved by separating philosophical writing from work intended to contribute to policy debates in bioethics. That has so far been my practice; its costs are rather high.

Here I have tried to link some serious philosophy with some consideration of institutions and practices. I have written with the thought that specific policies, practices and institutions can *illustrate* underlying philosophical questions and arguments, but equally in awareness that in other circumstances those principles and arguments might be illustrated by different policies, practices, or institutions. Since I have used a variety of bioethical illustrations of differing types, I have in any case aimed for sketches rather than for detailed blueprints. My illustrations are drawn mainly from the concerns of bioethics in the richer parts of the world, in which high-tech medicine and a culture of scientific research and biotechnological innovation flourish. More specifically, many of my illustrations are drawn from issues that have arisen in the UK, and to a lesser extent in the USA. Much writing in bioethics is dominated by examples drawn from the USA. There is, I believe, no harm (and possibly some gain) in extending the range of illustrations. I regret that it did not prove feasible to draw more illustrations from the practice of medicine outside the rich world. Had the lectures covered issues of justice in bioethics, the balance of illustrations would have been quite different.

I have had to be equally sketchy in discussing and referring to other work in bioethics. This is something of a relief. In bioethics massive footnoting often indicates insecurity rather than authority, and frequently directs the reader to sources that reiterate rather than establish central points. My practice – for which I offer no very complete justification – has been to cite quite selectively from philosophical, bioethical and other writing, and to provide a separate bibliography of institutional websites on which a range of relevant public documents can be found.

I have many to thank. They include the members of the Gifford Committee in the University of Edinburgh and John Frow who welcomed me back to the Institute for Advanced Studies in the Humanities in the University of Edinburgh; many Edinburgh philosophers, including Richard Holton, Rae Langton, Michael Menlowe and Stuart Sutherland; other Edinburgh friends and colleagues including Frances Dow, Duncan Forrester, Susan Manning, Paul McGuire, Charles Raab and Natasha and David Wilson; as well as members of Newnham College living in and near to Edinburgh. They all made giving the lectures more fun and more interesting than it would otherwise have been.

I owe a large debt to many Cambridge colleagues with whom I have discussed topics covered here across a number of years, and in particular to Martin Bobrow, Peter Lipton, Martin Richards and Marilyn Strathern. Since the lectures were delivered Stephen Buckle, Derek Burke, Norman Daniels, Peter Furness, Nicholas Harman, Patricia Hodgson, Andy Kuper and Tom Murray have helped me in many ways. Needless to say, remaining errors and implausibilities are my own contribution.

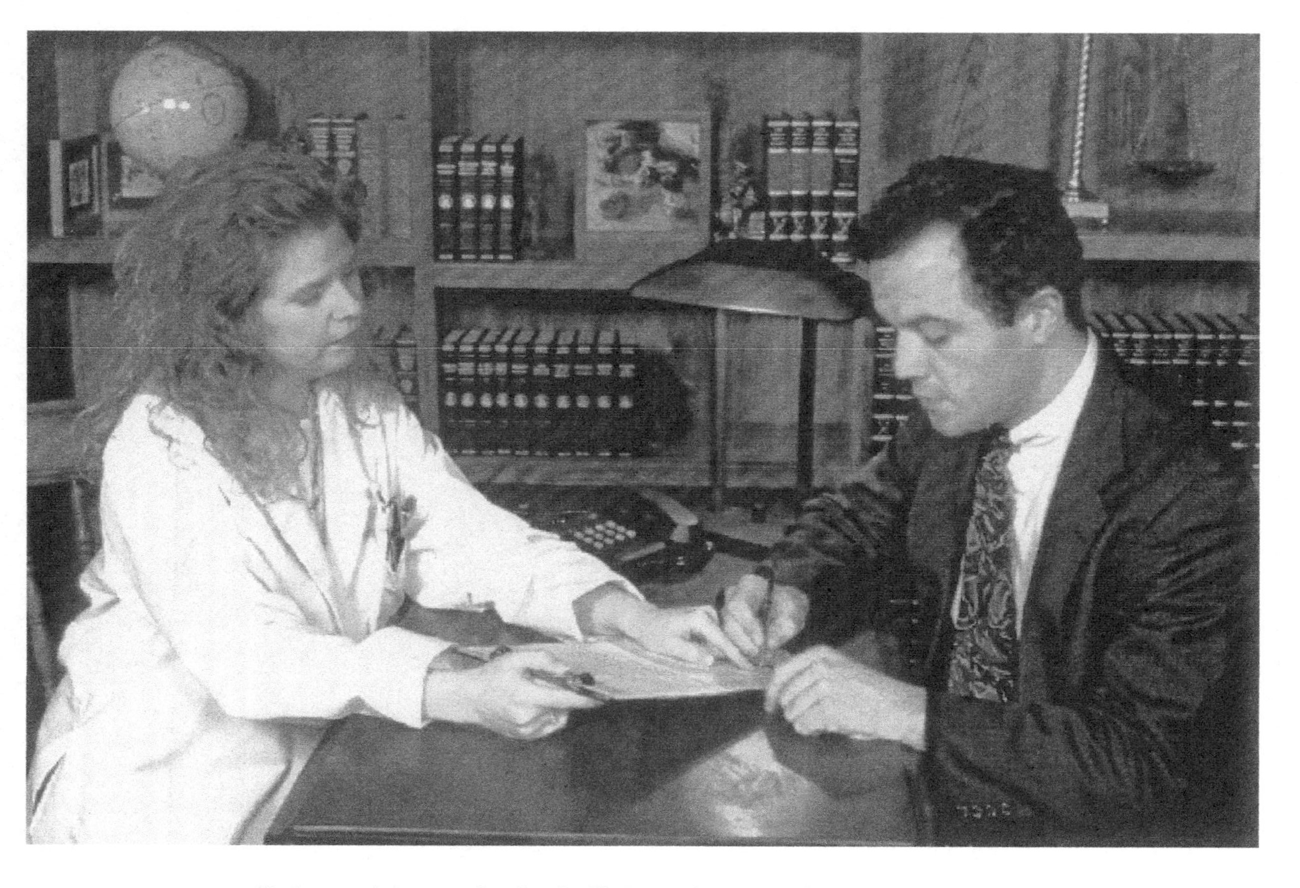

Patient and doctor: the ritual of informed consent. *Source:* ArtToday.com

CHAPTER ONE

Gaining autonomy and losing trust?

1.1 CONTEMPORARY BIOETHICS

Bioethics is not a discipline, nor even a new discipline; I doubt whether it will ever be a discipline. It has become a meeting ground for a number of disciplines, discourses and organisations concerned with ethical, legal and social questions raised by advances in medicine, science and biotechnology. The protagonists who debate and dispute on this ground include patients and environmentalists, scientists and journalists, politicians and campaigners and representatives of an array of civic and business interests, professions and academic disciplines. Much of the debate is new and contentious in content and flavour; some of it is alarming and some misleading.

The first occasion on which I can remember a discussion of bioethics – we did not then use the word, although it had been coined[1] – was in the mid-1970s at a meeting of philosophers, scientists and doctors in New York City. We were discussing genetically modified (GM) organisms: a topic of breathtaking novelty that was already hitting the headlines. Towards the end of the evening an elderly doctor remarked, with mild nostalgia, that when he had studied medical ethics as a student, things had been easier: the curriculum had covered referrals, confidentiality – and billing. Those simpler days are now very remote.

[1] The Kennedy Institute in Washington DC was founded in 1971 with the full name 'The Joseph and Rose Kennedy Institute for the Study of Human Reproduction and Bioethics'. See W. T. Reich, 'The Word 'Bioethics': Its Birth and the Legacies of Those Who Shaped It', *Kennedy Institute of Ethics Journal*, 4, 1994, 319–35.

During these years no themes have become more central in large parts of bioethics, and especially in medical ethics, than the importance of respecting individual rights and individual autonomy. These are now the dominant ethical ideas in many discussions of topics ranging from genetic testing to geriatric medicine, from psychiatry to *in vitro* fertilisation, from beginning to end of life problems, from medical innovation to medical futility, from heroic medicine to hospices. In writing on these and many other topics, much time and effort has gone into articulating and advancing various conceptions of respect for persons, and hence for patients, that centre on ensuring that their rights and their autonomy are respected. Respect for autonomy and for rights are often closely identified with medical practice that seeks individuals' informed consent to all medical treatment, medical research or disclosure of personal information, and so with major changes in the acceptable relationships between professionals and patients. Medical practice has moved away from paternalistic traditions, in which professionals were seen as the proper judges of patients' best interests. Increased recognition and respect for patients' rights and insistence on the ethical importance of securing their consent are now viewed as standard and obligatory ways of securing respect for patients' autonomy.[2]

Rights and autonomy have played a lesser, yet still a significant, part in other areas of bioethics, including even environmental ethics. For example, rights may be invoked in arguing for prohibitions on marketing unlabelled food products containing additives or GM crops or on adding chemicals to water supplies, with the thought that rights are violated where individuals cannot refuse, nor therefore choose, because they are kept in ignorance or unable to opt out. Agricultural regulations have been condemned as

[2] For a highly informative account of these changes, concentrated mainly on the US case, but with much that is relevant more widely, see Ruth Faden and Tom Beauchamp, *A History and Theory of Informed Consent*, Oxford University Press, 1986; for a sociological perspective see Paul Root Wolpe, 'The Triumph of Autonomy in American Bioethics: A Sociological View', in Raymond DeVries and Janardan Subedi, eds., *Bioethics and Society: Constructing the Ethical Enterprise*, Prentice-Hall, 1998, 38–59.

violating or as failing to protect animal rights, or farmers' rights to choose how to cultivate their land. Pollution controls have been attacked as violating the purported rights of individuals to conduct their lives and their businesses as they see fit.

We might expect the increasing attention paid to individual rights and to autonomy to have increased public trust in the ways in which medicine, science and biotechnology are practised and regulated. Greater rights and autonomy give individuals greater control over the ways they live and increase their capacities to resist others' demands and institutional pressures. Yet amid widespread and energetic efforts to respect persons and their autonomy and to improve regulatory structures, public trust in medicine, science and biotechnology has seemingly faltered. The loss of trust is a constant refrain in the claims of campaigning groups and in the press. In many developed countries, and particularly in the UK, there is evidence that mistrust of various professions, experts and of public authorities is quite widespread.[3]

This loss of trust is often ascribed to the supposed untrustworthiness of scientists and biotechnologists, even of doctors, and of those holders of public office who legislate for and regulate their activities. Medical professionals and regulators, politicians and civil servants, biotechnology companies and scientists, it is often suggested, pursue their own interests rather than those of patients or of the public. If true, these claims suggest that measures introduced (in part) to improve individual autonomy and to ensure that treatment and research do not proceed without informed consent have failed to secure trust, and may even have damaged trust. Perhaps this should not surprise us: increasing individual autonomy may increase the autonomy of those in positions of power, so adding to their opportunities for untrustworthy action and to others' reasons for mistrusting them. Perhaps reducing the autonomy of any agents and institutions who might act in untrustworthy ways would help to restore trust. Is some loss of trustworthiness and of trust an

[3] The MORI polls' website contains reports of numerous recent polls documenting lack and loss of public trust; see institutional bibliography (p. 205).

acceptable price for achieving greater respect for autonomy? Do we have to choose between respect for individual autonomy and relations of trust? None of these prospects would be particularly welcome: we prize both autonomy and trust. Yet can we have both?

1.2 MEDICAL ETHICS AND ENVIRONMENTAL ETHICS

The two principal domains of bioethics are medical ethics (broadly interpreted to include the ethics of bio-medical research) and environmental ethics. Autonomy and trust have played quite different roles in these two areas. The reasons behind these differences are instructive.

Much of medical ethics has concentrated on the individual patient, her rights and her autonomy; demands that medical professionals respect autonomy and rights have become a constant refrain. The implicit context of nearly all of this work is the medical system of a developed society with much hospital-based medicine. Topics such as the just distribution of health care within these medical systems, public health and global health distribution have been pushed to the margins in much of bioethics.[4] Perhaps these topics have been marginalised because individual autonomy is viewed as central to medical ethics.

Writing on environmental ethics has more often focused on public benefits and public harms. Here individual autonomy is quite often seen as a source of harms, and there has been a steadily increasing emphasis on the consequent need to limit individual autonomy. Standard examples of such controls include prohibitions on discharge of raw sewage or toxic chemicals, regulation of standards for vehicle emissions or building insulation and requirements for high safety standards in biotechnology. Contemporary discussions

[4] With notable exceptions. For an early example see Norman Daniels, *Just Health Care*, Cambridge University Press, 1985; a revised edition titled *Just Health* is forthcoming; also Thomas W. Pogge, 'Relational Conceptions of Justice: Responsibilities for Health Outcomes', in Sudhir Anand, Fabienne Peter and Amartya Sen, eds., *Health, Ethics, and Equity*, Clarendon Press, forthcoming. Questions of equity and fairness are generally more prominent in work on welfare, public health and health economics than they are in bioethics.

in environmental ethics seldom view the autonomous 'life-style' choices of individuals as adequate for protecting the environment. They increasingly highlight the importance of stewardship of the environment and argue that this requires public regulation and enforcement, sometimes international regulation and enforcement.

There are further and deeper reasons why individual autonomy has been less central in environmental than in medical ethics. Environmental ethics is fundamentally concerned with the treatment of life forms (above all of animals and plants), of groups and systems of life forms (such as ecosystems and populations), and with the importance of more abstract aspects of the environment such as species and the ozone layer, climate change and pollution. By and large, writing in environmental ethics has therefore tried to emphasise continuities between human and non-human parts of the natural world, and to claim for the latter some of the respect and concern traditionally thought important for the former. In claiming that the natural world is owed respect and concern, environmental ethicists have not viewed that world or its inhabitants as agents whose autonomy is to be fostered or whose consent to activities in which they are involved should be sought. Their ethical debates have therefore not been mainly concerned with agency and autonomy, with consent or anti-paternalism; rather their aim has been to detach notions such as rights, respect and concern from their historic association with conceptions of agency, persons and autonomy.

The distance between these two branches of bioethics is now diminishing. In part this is because several issues that link health and environmental concerns have become urgent. Discussions of GM crops, of food safety, of pollution and of animal welfare often link medical with environmental issues. The emergence of antibiotic-resistant strains of bacteria is a medical problem, for which poor agricultural practices may be partially responsible. Major environmental problems such as desertification, water shortages and air pollution all have serious health implications.

There is in any case more common theoretical ground between the two branches of bioethics than some suspect. Environmental ethics is, perforce, addressed to human agents: they are the only possible audience for its prescriptions and its arguments. It therefore has to build on the same assumptions about human agency that are basic to medical ethics. Although environmental ethics has often repudiated 'speciesism', and with it failures to take the claims – supposedly the rights – of various non-human parts of nature (especially of non-human animals) seriously, it is unavoidably every bit as anthropocentric in its view of the audience for ethical reasoning as any other bit of ethics.[5]

It is therefore not surprising that medical and environmental ethics have found a common language by focusing on rights. The language of rights permits convergence in the vocabularies of medical and environmental ethics by bracketing many questions about agency and obligation in favour of a primary focus on recipience and entitlement. Medical ethicists view human rights, among them patients' rights, as securing the right sort of respect for human agents and their autonomy. Environmental ethicists see the rights of animals, and even of other parts of the natural world such as plants and landscapes, ecosystems and species, as securing protection and respect for the non-human world.

Fundamentally the difference between these two parts of bioethics is not that one endeavour thinks agency important and that the other thinks it unimportant, but rather a focus on different objects of ethical concern, on the differing claims that these make on agents, and on the differing part that relationships between individuals play in the two domains. In medical ethics it has become standard to stress the *distinctiveness* of human capacities for agency, and to stress capacities for autonomy, and so to emphasise the special ethical concern and respect to be accorded to persons, including patients, and the special importance of human rights. In environmental ethics the *similarities* between human and non-human

[5] See Peter Singer, *Animal Liberation*, Jonathan Cape, 1976, for a critique of speciesism; for the relation of anthropocentrism to speciesism see Tim Hayward, 'Anthropocentrism: A Misunderstood Problem', *Environmental Values*, 6, 1997, 49–63 and Onora O'Neill, 'Environmental Values, Anthropocentrism and Speciesism', *Environmental Values*, 6, 1997a, 127–42.

parts of nature have been stressed: the normative claims, supposedly the rights, of humans and other primates,[6] of humans and all non-human animals,[7] of humans and non-human organisms in general have been compared, even equated. Most medical ethics is avowedly humanistic, but environmental ethicists regard humanism as an ethically unacceptable form of species preference (speciesism). They may even see human rights, let alone human autonomy, as problematic sources of harm or indifference to other living creatures or to the environment. Humanism is commonly seen as part of the *problem* rather than of the *solution* in environmental ethics. Nevertheless, both medical and environmental ethics can be addressed only to those who can reason, deliberate and act; both debates must take agency, and therefore human agency, seriously.

Since autonomy has played so much larger a role in medical than in environmental ethics, I shall mainly choose my illustrations from debates in medical ethics. However, I shall also introduce a limited range of examples from environmental ethics, in order to shed light both on reasons why the two parts of bioethics have diverged and on some ways in which public health issues have been marginalised in medical ethics.

1.3 TRUST IN THE RISK SOCIETY

Although discussions in medical ethics and environmental ethics have diverged in many other respects, both have recently encountered similar crises. In both areas agents and agencies have found it hard to establish and to maintain public trust in their action and policies. The crisis has been particularly marked in the UK, but is evident in many other rich and technically advanced societies.

The targets of public mistrust have been widely discussed across the last thirty years both in sociological discussions of the

[6] See Paola Cavalieri and Peter Singer, *The Great Ape Project: Equality beyond Humanity*, Fourth Estate, 1993.

[7] The best-known work is still Singer, *Animal Liberation*; but see also Stephen R. I. Clarke, *The Moral Status of Animals*, Oxford University Press, 1977; Peter Singer, *The Expanding Circle: Ethics and Sociobiology*, Clarendon, 1981.

'risk society'[8] and in the media. Leading sociologists have noted that many technical and social practices – prominently among them medicine, science and biotechnology – have become larger and more remote, and are seen as more laden with hidden risks, and that fears have multiplied with the globalisation of economic and technical processes. The fears and anxieties of 'risk societies' focus particularly on hazards introduced (or supposedly introduced) by high-tech medicine and genetic technologies, by nuclear installations and use of agrochemicals, by processed food and intrusive information technologies.

Yet it is open to doubt whether most people in the richer parts of the world encounter risks that they can do less to control than earlier generations could do to control risks they faced. Traditional hazards such as endemic tuberculosis or contaminated water supplies, food scarcity and fuel poverty were neither minimal nor controllable by those at risk from them in the recent past, and are neither minimal nor controllable for those who still face them in poorer societies today.[9] The claim that richer societies have become 'risk societies' is a claim not about levels of risk, but about changes in *perceptions of risk*, or at least in reported perceptions of risk. It is a claim about a supposedly widespread loss of confidence in the capacities of medical, scientific and technical progress to solve problems, and about a corresponding growth in reported anxiety and mistrust. These perceptions have currency among populations who in fact live longer and healthier lives than their predecessors enjoyed. Yet the claim about perceptions is accurate. In the UK, for example, MORI public opinion polls confirm that many

[8] Ulrich Beck, *Risk Society*, Sage, 1986; Piotr Sztompka, *Trust: A Sociological Theory*, Cambridge University Press, 1999.

[9] Other writers reject the doom-laden view that new technologies have increased risks. See Aaron Wildavsky, *Searching for Safety*, Transitions: Oxford University Press, 1988; also his 'If Claims of Harm from Technology are False, mostly False or Unproven What Does That Tell Us about Science?', chapter 10 in Peter Berger *et al.*, eds., *Health, Lifestyle and Environment*, Social Affairs Unit. See also John Adams, *Risk*, UCL Press, esp. pp. 179–80, and many of the papers in Julian Morris, ed., *Rethinking Risk and the Precautionary Principle*, Butterworth Heinemann, 2000.

members of the public now claim to distrust numerous groups and professionals to tell the truth about medical, scientific and environmental issues.[10]

UK media accounts of these polls and the public attitudes they sample report that the public do not trust science, industry or politicians. There is also a limited amount of evidence that perceived lack of trust is expressed in action: there are sporadic environmental protests and demonstrations, there is widespread public refusal to buy GM foods and quite a lot of people buy 'alternative' medicines (despite the fact that most have been tested neither for safety nor for efficacy). Yet there is also a great deal of evidence of action that suggests that the public do *not* mistrust scientists, industry or politicians any more than they mistrust others, and that they do not (for the most part) lose trust in entire professions or industries when they become aware of untrustworthy behaviour by a few.[11] Despite some highly publicised professional failures and crimes, there is good evidence that the public continue to place trust not only in doctors, but also in the scientists who develop new medicines, in the industries that produce them and in the regulators who ensure safety standards. Loss of trust, it seems, is often reported by people who continue to place their trust in others; reported perceptions about trust are not mirrored in the ways in which people actually place their trust.

[10] For MORI polls on GMO, see institutional bibliography. Other studies have recorded slightly varying rankings: see L. J. Frewer, C. Howard, D. Heddereley and R. Shepherd, 'What Determines Trust in Information about Food-Related Risks? Underlying Social Constructs', in Ragnar Löfstedt and Lynn Frewer, *Risk and Modern Society*, Earth Scan, 1998, 193–212, see esp. table on p. 198, in which the least trusted information sources, in order, are tabloid newspapers, MPs, ministers, ministries and personal friends(!) and the most trusted are university scientists, medical doctors, consumer organisations, television documentaries and government scientists.

[11] In the UK cases of concern about failures in medical practice are documented in the 2001 Redfern Report on events at Alder Hey hospital and the 1995 Kennedy Report on events at the Bristol Heart Unit. Since the publication of the Redfern Report, the British Medical Association (BMA) has commissioned a poll from MORI, which showed that the public still retains greater trust in doctors than in any other group. See institutional bibliography for all sources, and especially MORI/BMA 2001 on the MORI website.

Claims about mistrust and its practical implications are nevertheless very prominent in public debate. Some influential voices advocate strong and barely coherent interpretations of the famous (if elusive) *precautionary principle*. They suggest, for example, that all and any innovations that may harm the environment should be prohibited, regardless of likely benefits: yet very few changes are guaranteed to have *no* bad effects; even fewer can be guaranteed in advance to be harm-free; and even the status quo (as some of the same voices complain) may have bad effects – so presumably should also not continue.[12] But what does the precautionary principle prescribe when both change and the status quo are judged wrong? There are also many demands for impractical levels of safety and success in medical practice and environmental standards, such as claims that everybody should receive 'the best' treatment: possible only where zero variation of treatment is guaranteed. There are demands that no traces of substances that pollute in large quantities should be permitted in water or food (salt?).[13] There are even occasional demands for a supposed (but literally speaking incoherent) 'right to health', a fantasy that overlooks the fact no human action can secure health for all, so that there can be no human obligation to do so, and hence no right to health. These excessive and unthought-through demands are evidence of a culture in which trust is besieged. Debate is often shrill and hectoring. A culture of blame and accusation is widespread, both in the media and in the literature of campaigning organisations, where fingers are pointed variously at government, at scientists and at business.[14]

[12] For a survey of stronger and weaker interpretations of the principle see Julian Morris, 'Defining the Precautionary Principle' in Julian Morris, ed., *Rethinking Risk and the Precautionary Principle*, Butterworth Heinemann, 2000, 1–21; and Aaron Wildavsky, 'Trial and Error versus Trial without Error', in Morris, ed., 2000, 22–45.

[13] For example the most recent text of the World Medical Association, Declaration of Helsinki benchmarks requirements in medical research by reference to 'best' treatment; see institutional bibliography.

[14] For a useful case study see Parliamentary Office of Science and Technology (POST), *The 'Great GM Food Debate': A Survey of Media Coverage in the First Half of 1999*, 138, May 2000; for suggestive examples see Richard North, 'Science and the Campaigners', *Economic Affairs*, 2000, 27–34.

This looming atmosphere of distrust has arisen amid, and co-habits with, great and well-publicised advances in medicine and the life sciences, in biotechnology and in protection of the environment. Scientific success and reduction of risks to life, health and the environment are manifest not only in research, but also in the application of research to medical practice and environmental protection. Life expectancy has risen and is still rising in the richer world, and also in many (but not all) parts of the poorer world. Medical care has been improving, and many serious health problems are now ones that individuals can address for themselves, for example by stopping smoking or drug use, or by losing weight or exercising more. Even the much criticised – but also much loved – National Health Service (NHS) progresses towards evidence-based medicine. Equally in environmental matters, in the UK and in some other richer countries, air and water are becoming cleaner; greener technologies and energy savings are pursued; agricultural practices that cause environmental harm are being reduced; biodiversity is monitored and the news on biodiversity and wildlife is quite often encouraging.[15] There is even an increasing public recognition that environmental standards matter and must be paid for. In short, reported public trust in science and even in medicine has faltered *despite* successes, *despite* increased efforts to respect persons and their rights, *despite* stronger regulation to protect the environment and *despite* the fact that environmental concerns are taken far more seriously than they were a few years ago.

Taken at face value, the mismatch between increasing advances in safety standards and environmental concern and declining reported trust is strange. Why should trust be declining at a time when reasons for trusting have apparently grown? There could be various good explanations for this surprising fact. For example, some ascribe the current culture of mistrust to the public's lack of scientific education (remedy: improve public understanding of science), and others ascribe it to the poor communication

[15] Tony Gilland, 'Precaution, GM Crops and Farmland Birds', in Julian Morris, ed., *Rethinking Risk and the Precautionary Principle*, Butterworth Heinemann, 2000, 60–83.

skills of doctors and scientists (remedy: teach doctors and scientists to communicate better), or to deeper and persisting conflicts of interest. I shall comment on some of these diagnoses in later chapters. However, I want first to consider the more fundamental difference between perceiving others as trustworthy and actively placing trust.

1.4 JUDGING RELIABILITY AND PLACING TRUST

Loss of trust has become a major issue in public debate, but there has been less discussion of trust and loss of trust in bioethics, or in ethics more generally, than one might have expected. Trust has been a major theme in sociology, but only a minor theme in ethics. In consequence a large amount of discussion of trust focuses on empirical studies of perception of others as trustworthy or untrustworthy, and rather little addresses the practical demands of placing trust. The topics are connected, but they are not the same. The connection is that those who see their world as a 'risk society' often find placing trust problematic: but it does not follow that they do not place trust, or even that they place no trust in those whom they claim to think untrustworthy.

Just as total scepticism would produce total paralysis of belief, and is untenable in practice, so total inability to place trust would produce total paralysis of action, and is untenable in practice. In practice we have to take a view and to place our trust in some others for some purposes. Where people perceive others as untrustworthy they may place their trust capriciously and anxiously, veering between trusting qualified doctors and trusting unregulated alternative practitioners, between trusting scientific claims and trusting those of alternative, greenish or counter-cultural campaigners, or modish therapies and diets, between trusting established technologies and medicines and trusting untested or exotic technologies and products. But they do not refuse to trust.

The thought that anyone who sees others as untrustworthy can avoid placing trust is unconvincing. In trusting others to do or refrain from action of a certain sort we do not assume any guarantee that they will live up to that trust.[16] Trust is not a response to certainty about others' future action. On the contrary, trust is needed precisely when and because we lack certainty about others' future action: it is redundant when action or outcomes are guaranteed. That is why we find it hard, as well as important, to try to place trust reasonably rather than foolishly.

Usually we place trust in others only with respect to a specific range of action, often for action for which they have explicit responsibility. A patient may trust her doctor to act in her best interests in deciding on her treatment, but might not trust him to drive safely. A parent may trust a schoolteacher to teach his child, but not to look after his money or to diagnose an illness. A householder may trust a water company to provide safe tap water, but not to deliver the groceries. However, in other cases trust is unrelated to role. We cannot avoid trusting strangers in many matters, like driving on the correct side of the road or giving what they take to be reliable rather than invented information when asked. And we cannot avoid placing many different sorts of trust in others with whom we have close and complex relationships. In personal relationships trust is often reciprocal and may be given for a very wide range of action.

When we place trust in others, we do not usually trust or even expect them to have our interests entirely at heart, let alone to place our interests ahead of all other concerns. One of the rare, and influential, accounts of the ethics of trust, proposed by Annette Baier, suggests that when we place trust in others we not merely rely on them, but rely on them having at least minimal good will towards us:

> Reasonable trust will require good grounds for . . . confidence in another's good will, or at least the absence of grounds for expecting their ill will or indifference.[17]

[16] Sztompka, *Trust*; Annette Baier, 'Trust and Antitrust', *Ethics*, 96, 1986, 231–60.

[17] Annette Baier, 'Trust and Antitrust', 235. For a similar view, emphasising the importance of good will for trust see Karen Jones, 'Trust as an Affective Attitude', *Ethics*, 107, 1996, 4–25. For an insightful and in my view more plausible analysis of trust,

But this is often not the case. Our trust in individuals and in institutions, in officials and in professionals, does not (fortunately!) rest on the thought that they have good will towards us. The thought that placing trust requires good will has a context (at most) in personal relationships – and perhaps not in all of those.

We therefore need a broader view of placing trust, that takes account of the fact that we often trust others to play by the rules, achieve required standards, do something properly without the slightest assumption that they have any good will towards us. Sometimes we may know that good will is lacking, and yet trust. A patient may know that a doctor finds him particularly irritating and bears him little good will, and yet trust the doctor to exercise proper professional judgement. Most of us trust the safety of ordinary medicines without knowing much, if anything, about the procedures for safety and efficacy testing to which they have been subjected, or about the companies and regulatory bodies responsible for these procedures, let alone assuming that these companies and regulatory bodies have good will towards us. What is the basis of placing trust when good will does not enter the picture?

It is often thought that we place trust in others because they have proved reliable, and that we withdraw trust from them because they have proved unreliable. Views of others' reliability are useful in placing trust, but they are neither necessary nor sufficient for doing so. In judging that someone is reliable we look to their past performance; in placing trust in them we commit ourselves to relying on their future performance. We can see that knowledge of others' reliability is not necessary for trust by the fact that we can place trust in someone with an indifferent record for reliability, or continue to place trust in others in the face of some past unreliability. Many daily relationships of trust survive a good deal of failure and unreliability; we commonly regard those who withdraw trust after a single lapse (or even after sporadic minor lapses) as excessively suspicious. Proven reliability may be nice, but it is not necessary for placing trust. Equally, we can see that reliability

see Richard Holton 'Deciding to Trust, Coming to Believe', *Australasian Journal of Philosophy*, 72, 1994, 63–76.

is not sufficient for placing trust, both because trust is not directed to natural processes (however reliable) but only to other agents, and because reliable agents are not always trusted.

In judging reliability we draw largely on evidence of past performance; in placing trust we look to the future, and evidence of past conduct is only one of the factors we commonly consider. We expect competent persons to converge in judgements of reliability if they have access to the same evidence; we do not expect the same convergence in placing of trust. If we imagined that placing trust was dictated entirely by another's past record for reliability, we could make no sense of many significant decisions to place trust in others. We could not understand amnesties, or reconciliation, or forgiveness, or confidence building: all are instances of placing trust despite poor evidence of past reliability. Placing trust is not dictated by what has happened: it is given, built and conferred, refused and withdrawn, in ways that often go beyond or fall short of that evidence.

Nevertheless the most common explanation for refusal to place trust is that it is a reasonable response to prior untrustworthiness or unreliability, and correspondingly that trust is a proper response to prior trustworthiness or reliability. For example, distrust of medicine, science and biotechnology is often said to be justified by past action or inaction that has damaged public interests or abused public trust during the last fifty years. Regularly cited examples include the incautious introduction of DDT, the unregulated use of organophosphates; and the building of nuclear power plants without adequate plans for nuclear fuel reprocessing. More recently in the UK mistrust is said to have been caused by poor government handling of the emergence of BSE in cattle, by the one-sided attitude to the introduction of GM crops taken by Monsanto and some others, by worries created by geographically erratic availability of certain forms of medical treatment and by some highly publicised cases of professional malpractice. All of these factors, and many others, may offer some reasons for the public to judge some of those who practise medicine, science and biotechnology unreliable. However those judgements about past reliability invariably underdetermine their decisions about where to place their trust.

This is not as irrational as it may at first seem. Judgements of reliability are in any case often based on limited and inconclusive evidence. Well-publicised cases of untrustworthy action by professional and office holders offer very incomplete reasons for judging all other professionals or office holders, or even the same ones in a different situation, untrustworthy. In many cases the available evidence is sufficiently porous for agents to find it reasonable either to place or to refuse trust – or to claim to mistrust while in practice placing trust.

If all claims not to trust medicine, science and biotechnology were based on comprehensive evidence of systematic unreliability, past performance would present an extreme challenge to placing further trust. But claims that others are untrustworthy of the sort that are now so common often reflect very incomplete evidence. I shall explore a range of thoughts about sources of claims to mistrust medicine, science and biotechnology. Might it be the case that mistrust sometimes arises even without any knowledge of (significant or widespread) prior failure of reliability, for example because it is too hard to distinguish accurate information from misinformation and disinformation, so too hard to place trust reasonably? Might it sometimes arise from very procedures by which we try to make medical and scientific practice more accountable, and in particular from ways in which we have tried to combine respect for the autonomy of patients and of members of the public with regulatory protection? Or could the very conceptions of autonomy and of respecting autonomy, that have been at the heart of so many policies for regulating medicine, science and biotechnology, threaten the maintenance and creation of trust? Is loss of trust perhaps the price of increasing autonomy? Must we choose between respect for autonomy and relations of trust?

1.5 TRUST AND AUTONOMY IN MEDICAL ETHICS

Answers to all of these questions are complicated because various conceptions of autonomy and of trust are in play, between which

I hope to distinguish. In doing so I shall try to say something about various conceptions of each, and to trace some of their relations to other ideas that are prominent in contemporary bioethics, such as those of respect for persons, informed consent and certain human rights.

I hope to show that some conceptions of autonomy and of trust are compatible, and even mutually supporting. It will not, of course, follow that we must adopt these conceptions of autonomy and of trust. We may find reason to prefer others. However, if we rely on conceptions of autonomy and of trust that cannot be reconciled, then we cannot have both. Correspondingly, if we would like to find a way of enjoying both autonomy and trust we must first find conceptions of each that can be reconciled.

I shall begin the inquiry by posing some intuitive questions about the relation of trust to autonomy within medical ethics, for it is in medical ethics that some of the strongest claims have been made both on behalf of trust and on behalf of autonomy. If we think back into the past, and look to that famous prototype of all professional relationships, the doctor–patient relationship, we have a paradigm of a relationship of trust. The patient approaches the doctor knowing that the doctor is bound as a matter of professional oath and integrity to act in the patient's best interests, even that the doctor stands at risk of disgrace or disqualification for serious failure in this regard. Although there are always contractual and financial arrangements linking doctor and patient, or doctors and the institutions that organise medical care and employ them, the doctor–patient relationship is supposed to trump any considerations of self-interest and gain. It is a professional relationship that is supposed to be disinterested, long-lasting, intimate and trusting. The image in the frontispiece of this book can be seen as depicting a trusting, traditional doctor–patient relationship, one-to-one, indeed face-to-face, set in the confidential confines of a professional office.

This traditional model of the trusting doctor–patient relationship has been subject to multiple criticisms for many years. Traditional doctor–patient relationships, it has been said on countless occasions, have in fact nearly always been based on asymmetric knowledge and power. They institutionalise opportunities for abuse of trust. Doctor–patient relationships were viewed as relationships of trust only because a paternalistic view of medicine was assumed, in which the dependence of patients on professionals was generally accepted. The traditional doctor–patient relationship, so its critics claim, may have been one of trust, but not of reasonable trust. Rather, they claimed, patients who placed trust in their doctors were like children who initially must trust their parents blindly. Such trust was based largely on the lack of any alternative, and on inability to discriminate between well-placed and misplaced trust.

If there was one point of agreement about necessary change in the early years of contemporary medical ethics, it was that this traditional, paternalistic conception of the doctor–patient relationship was defective, and could not provide an adequate context for reasonable trust. A more adequate basis for trust required patients who were on a more equal footing with professionals, and this meant that they would have to be better informed and less dependent. The older assumption that relations of trust are in themselves enough to safeguard a weaker, dependent party was increasingly dismissed as naive. The only trust that is well placed is given by those who understand what is proposed, and who are in a position to refuse or choose in the light of that understanding. We can look at the same image with a less innocent eye, and see it as raising all these questions about the traditional doctor–patient relationship. In this second way of seeing the picture the doctor dominates: the white coat and intimidating office are symbols of her professional authority; the patient's anxious and discontented expression reveals how little this is a relationship of trust.

These considerations lie behind many discussions of supposedly better models of the doctor–patient relationship, in which patients are thought of as equal partners in their treatment, in which treatment is given only with the informed consent of patients, in which patient satisfaction is an important indicator of professional adequacy, in which patients are variously seen as consumers, as informed adults and are not infantilised or treated paternalistically and in which the power of doctors is curbed.[18] In this more sophisticated approach to trust, autonomy is seen as a precondition of genuine trust. Here, as one writer puts it, 'informed consent is the modern clinical ritual of trust',[19] a ritual of trust that embeds it in properly institutionalised respect for patient autonomy. So we can also read the image in the frontispiece in a third, more optimistic, way as combining patient autonomy with mutual trust in the new, recommended, respecting way. What we now see is a relationship between equals: the patient too is a professional, dressed in a suit and sitting like an equal at the desk; the patient has heard a full explanation and is being offered a consent form; he is deciding whether to give his fully informed consent. Trust is properly combined with patient autonomy.

This revised model of doctor–patient interaction demands more than a simple change of attitude on the part of doctors, or of patients. It also requires huge changes in the terms and conditions of medical practice and ways of ensuring that treatment is given only where patients have consented. Informed consent has not always been so central to doctor–patient relationships, which were traditionally grounded in doctors' duties not to harm and to benefit. Informed consent came to be seen as increasingly important in part because of legal developments, especially in the USA, and in part because of its significance for research on human subjects, and the dire abuse of research subjects by Nazi doctors. The first principle of the Nuremberg Doctors' Code of 1947 states emphatically that subjects' consent must be 'voluntary, competent, informed and comprehensive'.[20] Only later did the thought emerge clearly that consent was also central to clinical practice, and that patient autonomy or self-determination should not be subordinated to doctors'

[18] R.A. Hope and K.W.M. Fulford, 'Medical Education: Patients, Principles, Practice Skills', in R. Gillon, ed., *Principles of Health Care Ethics*, John Wiley & Sons, 1993.

[19] Wolpe, 'The Triumph of Autonomy', 48.

[20] See Faden and Beauchamp, *A History and Theory of Informed Consent*; Ulrich Tröhler and Stella Reiter-Theil, *Ethics Codes in Medicine: Foundations and Achievements of Codification Since 1947*, Ashgate; Lori B. Andrews, 'Informed Consent Statutes and the Decision-Making Process', *Journal of Legal Medicine*, 30, 163–217; World Medical Association, Declaration of Helsinki, 2000; see institutional bibliography.

commitments to act for their patients' benefit or best interest. Yet despite the enormous stress laid on individual autonomy and patient rights in recent years, this heightened concern for patient autonomy does not extend throughout medicine: public health, and the treatment of those unable to consent are major domains of medical practice that cannot easily be subjected to requirements of respecting autonomy and securing informed consent.[21]

From the patient's point of view, however, the most evident change in medical practice of recent decades may be loss of a context of trust rather than any growth of autonomy. He or she now faces not a known and trusted face, but teams of professionals who are neither names nor faces, but as the title of one book aptly put it, *strangers at the bedside.*[22] These strangers have access to large amounts of information that patients give them in confidence. Yet to their patients they remain strangers – powerful strangers. They are the functionaries of medical institutions whose structures are opaque to most patients, although supposedly designed to secure their best interest, to preserve confidentiality and to respect privacy. Seen 'from the patient's point of view every development in the post World War II period distanced the physician and the hospital from the patient, disrupting social connection and severing the bonds of trust'.[23]

From the practitioner's point of view, too, the situation has losses as well as gains. The simplicities of the Hippocratic oath and of other older professional codes have been replaced by far more complex professional codes, by more formal certification of competence to perform specific medical interventions, by enormous increases in requirements for keeping records and by many exacting forms of professional accountability.[24] In medicine, as in most

[21] See chapter 2. The marginalisation of these topics may reflect their poor fit with the popular ideal of patient autonomy.

[22] David J. Rothman, *Strangers at the Bedside: A History of How Law and Ethics Transformed Medical Decision-Making*, Basic Books, 1991. Rosamond Rhodes and James J. Strain, 'Trust and Transforming Healthcare Institutions', *Cambridge Journal of Healthcare Ethics*, 9, 2000, 205–17.

[23] Rothman, *Strangers at the Bedside.*

[24] Nigel G.E. Harris 'Professional Codes and Kantian Duties', in Ruth Chadwick, ed., *Ethics and the Professions*, Amesbury, 104–15. See chapter 6 below.

other forms of professional life and public service, an 'audit society' has emerged.[25] The doctor now faces the patient knowing that he or she must comply with explicit standards and codes, that many aspects of medical practice are regulated, that compliance is monitored and that patients who are not properly treated may complain – or even sue.

These new relationships may live up to their billing by replacing traditional forms of trust with a new and better basis for trust. The new structures may provide reasons for patients to trust even if they do not know their doctors personally, and do not understand the details of the rules and codes that constrain doctors' action. Supposedly they can feel reassured that the power of doctors is now duly regulated and constrained, that doctors will act with due respect and that they can seek redress where doctors fail. Although traditional trust has vanished with the contexts in which it arose, a more acceptable basis for reasonable trust has been secured, which anchors it in professional respect for patients' rights. Supposedly the ideals of trust and autonomy have been reshaped and are now compatible.

1.6 VARIETIES OF AUTONOMY

To judge whether autonomy and trust as now construed are indeed compatible, we need a rather clearer view of autonomy. This is not easily acquired. Gerald Dworkin began a book on autonomy by listing about a dozen distinct understandings of the notion. He suggested that it has been variously equated with

> Liberty (positive or negative) . . . dignity, integrity, individuality, independence, responsibility and self-knowledge . . . self-assertion . . . critical reflection . . . freedom from obligation . . . absence of external causation . . . and knowledge of one's own interests.[26]

[25] Michael Power, *The Audit Explosion*, Demos, 1994 and *The Audit Society: Rituals of Verification*, Oxford University Press, 1994.

[26] Gerald Dworkin, *The Theory and Practice of Autonomy*, Cambridge University Press, 1988, 6. See also his 'The Concept of Autonomy', in John Christman, ed., *The Inner Citadel: Essays on Individual Autonomy*, Oxford University Press, 1989, 54–76, esp. p. 54

Ruth Faden and Thomas Beauchamp suggest in their immensely interesting and useful book on informed consent that autonomy may also be identified with

privacy, voluntariness, self-mastery, choosing freely, choosing one's own moral position and accepting responsibility for one's choices.[27]

I have no idea whether these lists include all possibilities, but suspect that they do not: we might, for example, add the terms 'self-control' and 'self-determination'.

Dworkin thinks that

The only features that are held constant from one author to another are that autonomy is a feature of persons and that is a desirable quality to have.[28]

This is hardly an exacting claim, yet I doubt whether it is correct on either point. There are a lot of writers – they include many feminists, virtue ethicists and communitarians – who doubt whether autonomy is always of value. There are others, including various determinists, behaviourists and structuralists, who think that it is an illusion. There are also defenders of one or another conception of autonomy who think that it is not a feature of persons, either because they think that it is a feature of some but not of all persons, or because they think that it pertains not to persons but (for example) to the will, or to certain actions, or to certain principles, rather than to persons.[29] However, Dworkin's list provides a very valuable starting point for thinking about autonomy in bioethics,

and Thomas E. Hill Jr., 'The Kantian Conception of Autonomy', in his *Dignity and Practical Reason in Kant's Moral Theory*, Cornell University Press, 1992, 76–96, who begins the article with the observation 'Autonomy is a central concept in contemporary moral debates as well as in the discussion of Kant; but the only thing that seems completely clear about autonomy in these contexts is that it means different things to different writers.'

27 Faden and Beauchamp, *The History and Theory of Informed Consent*, 7.

28 Dworkin, *The Theory and Practice of Autonomy*, 6.

29 Thomas E. Hill, Jr., points out in 'The Kantian Conception of Autonomy' that Kant never predicates autonomy of persons, but only of principles and willings; Mill predicates autonomy of states, but not of persons. See chapter 2.

because it shows how many different notions may be intended, and how multiply ambiguous claims about the value of autonomy may be.

Despite this proliferation of conceptions of autonomy, there is probably more agreement about it in contemporary bioethics than elsewhere. In bioethics, and in particular in medical ethics, autonomy has most often been understood as a feature of individual persons. It is generally seen as a matter of *independence*, or at least as a *capacity for independent decisions and action*. This conception of individual autonomy sees it as *relational*: autonomy is always autonomy from something; as *selective*: individuals may be independent in some matters but not in others; and as *graduated*: some individuals may have greater and others lesser degrees of independence.

Although many protagonists of autonomy in bioethics claim to derive their moral reasoning either from Mill or from Kant (in chapters 2 and 4 I shall discuss these common thoughts about provenance), it seems to me likely that prevailing views of autonomy as independence owe as much or more to twentieth-century conceptions of character and individual psychology and to studies of moral development than they do to older traditions of moral philosophy. If we cast our minds back to the early post-Second World War period, we find intense interest in the fact that some people achieve more independence in the face of catastrophe than do others. In a world in which collaboration with and resistance to evil-doing had been of immense importance, the psychological differences between those who had collaborated and conformed and those who had resisted and stood up to be counted were of great ethical importance. The theme was fundamental to Adorno's *The Authoritarian Personality*,[30] which contrasted the deference of those with authoritarian personalities with the independence shown by those with democratic personalities. Similar thoughts were prominent in writing on perpetrators and victims in the concentration camps, for example in the work of Bruno Bettelheim and Primo Levi, who contrasted those whose capacities for

[30] Theodore W. Adorno, *The Authoritarian Personality*, Harper & Bros, 1950.

independent and ethical action failed in the death camps with those who survived as persons. The contrast was also central to the notorious Milgram experiments, in which volunteers were invited to punish experimental subjects who failed to learn simple tasks by administering electric shocks. Some deferential and conforming subjects proved willing to inflict high levels of pain (in fact they administered none, since the pain was mimed by actors colluding with the experimenters) simply because they had been told to do so.[31]

Twentieth-century studies of moral development in children also often focused on conceptions of autonomy as independence. In the 1930s Piaget's pioneering *Moral Judgement of the Child*[32] distinguished the immaturity of children who thought of moral requirements as a matter of obeying immutable rules, from the greater maturity of those who reviewed and revised rules. Similar distinctions were central to the cross-cultural studies of moral development undertaken by Lawrence Kohlberg, who also identified moral maturity with individual autonomy in choosing and criticising rules.[33]

It is, I think, no great mystery that autonomy should have been understood as a matter of individual independence in and beyond bioethics for some decades. Yet if autonomy is a matter of independence, it is very easy to see why it bears hard on relations of trust. Independent people may be self-centred, selfish, lacking in fellow-feeling or solidarity with others – in short, the very people in whom one would have least reason to place trust and who might encourage a culture of mistrust. Alcibiades was splendidly autonomous, and betrayed all the trust placed in him. Once we interpret autonomy *simply* as independence from others, or from others' views or their preferences, the tension between autonomy

[31] Stanley Milgram, *Obedience to Authority: An Experimental View*, Tavistock Publications, 1974.

[32] Jean Piaget, *The Moral Judgement of the Child*, Penguin, 1977.

[33] Lawrence Kohlberg, *The Philosophy of Moral Development*, Harper & Row, 1981. Kohlberg was criticised in turn in the 1980s and 1990s by Carol Gilligan, on the grounds that he identifies greater moral maturity with autonomous rule-making rather than with building relationships with others, so uncritically presupposing an allegedly 'male' view of what is ethically important; see Carole Gilligan, *In A Different Voice: Psychological Theory and Women's Dependence*, Harvard University Press, 1982; 2nd edn., 1993.

and trust is unsurprising. Trust is most readily placed in others whom we can rely on to take our interests into account, to fulfil their roles, to keep their parts in bargains. Individual autonomy is most readily expressed when we are least constrained by others and their expectations. Trust flourishes between those who are linked to one another; individual autonomy flourishes where everyone has 'space' to do their own thing.[34] Trust belongs with relationships and (mutual) obligations; individual autonomy with rights and adversarial claims.

If we are worried about loss of trust, we may wonder whether and why individual autonomy should now be so much admired. Surely independence is admirable in some cases and contexts, but not in others? One of my students illustrated this rather well at about the time that I first encountered bioethics. She joined a group of male students in welcoming spring weather to New York City, only to have the Columbia University student newspaper publish a photograph of them streaking across Broadway. I asked her why she had done it, and she told me that she felt that she had finally proved that she was autonomous. It was clear enough that her action was independent in some ways, although possibly not in others (did she not defer to male initiative?). She may well have been thinking that she had now shown herself independent of her parents, or of social conventions. However, this sort of independence doesn't invariably have merit. Independent action can be important or trivial, heroic or brutal, helpful or selfish, admired or distressing to others. If we view individual autonomy as mere, sheer independence, its merits will be highly variable. We would need some deeper set of reasons, or a deeper conception of autonomy, to explain why individual autonomy is ethically important. The fact that individual independence in the face of evil, or of temptation, is admirable does not show that individual independence in the face of others' needs, or in the context of family or professional relationships will be good or right. Presumably there has to be something over and above mere, sheer independence

[34] See Thomas H. Murray, *The Worth of a Child*, University of California Press, 1996.

that has made appeals to individual autonomy so attractive and ubiquitous in contemporary bioethics.

Some sociologists of medicine have suggested that the appeal of autonomy, understood as individual independence, in medical ethics is that it gives only the *illusion* of challenging professional authority, while *in fact* leaving that authority largely intact. The autonomous patient is not actually going to be allowed to determine his or her own treatment. He or she is only going be allowed to accept or refuse treatment proposed by professionals: the cash value of what is termed 'patient autonomy' is a right to refuse treatment that is offered, a right that is costly to exercise where there are few or no other options of treatment.[35] Undoubtedly such rights are of great value: they are what stand between patients and coerced treatment, and there are good reasons for taking the greatest care about any use of coercion in medicine. Nevertheless this right does not secure any distinctive form of individual autonomy or independence. Anyone who doubts this has only to consider what happens to a patient who demands treatment not available in a particular context. A limited right to refuse does not require capacities for independent, reflective choice, but it may be used to transfer formal responsibility for choice of treatment (and even for failure of treatment) to patients – who may yet feel quite powerless. Patient complaints can be rebutted with the claim that *volenti non fit iniuria*, and the power of health systems and professionals will not be greatly reduced since they will always control the agenda by determining what is to be offered. On this view what is misleadingly spoken of as 'patient autonomy' masks the fact that the patient's role is only to say 'yes' – or to do without treatment.

If we bring these thoughts to a final reading of the image in the frontispiece we reach a more suspicious reading of 'the modern clinical ritual of trust'. On this fourth reading the doctor has set out the options, and is now *telling* the patient to sign, and where

35 A commercially based medical system, as in the USA, may offer more options for 'shopping around' physicians. However options will still be limited both by the fact that professional judgement is not at the beck and call of patients, and by the typical financial constraints placed by systems of 'managed care'.

to sign. Look at that emphatically pointing finger! The patient is being told firmly that he is autonomous, that he is an equal partner in treatment and that he is about to give his free consent: but the reality, as his pained expression suggests, is quite different. This scene illustrates neither traditional trustworthiness and trust, nor their failure, nor newer and better grounded trust combined with respect and autonomy: it illustrates a simulacrum of autonomy – and a simulacrum of trust, just as the orderly office with its shelves of untouched, gold lettered volumes (evidently bought by the yard!) depicts a simulacrum of the real settings of professional life.

I do not, of course, want to suggest that patients' rights to refuse are unimportant. But where options are few, where cognitive and decision-making capacities are limited, procedures of informed consent may become a burden or a ritual, and ideas of 'patient autonomy' may seem more inflationary than liberating. If autonomy is really fundamental to bioethics, we need an ethically more convincing account of autonomy. I hope to provide that account in later chapters.

CHAPTER TWO

Autonomy, individuality and consent

2.1 THE ORIGINS OF INDIVIDUAL AUTONOMY

Most contemporary accounts of autonomy see it as a form of *independence*. Independence is relational: it is independence from something or other. So we may reasonably ask from what autonomous persons and autonomous action are independent. Many sorts of action are independent in some respects but dependent in others, and some sorts of independence do nothing to show that an action is right or valuable. Some independent action is spontaneous, disciplined, altruistic and even heroic; some is self-centred, pig-headed, impulsive, random, ignorant, out of control and regrettable or unacceptable for these and many other reasons.

A standard response to this deflationary thought is to see autonomous choosing as independent in some special and desirable way. Individual autonomy is not a matter of mere, sheer independence, of the sort praised by pop-existentialists, or aspired to by my streaking student. Whatever else people think about individual or personal autonomy, they do not equate it with mere choice.[1] Yet they disagree spectacularly about what makes some choosing autonomous, or more autonomous than other choosing, and other choosing less autonomous. A look at some current conceptions of individual autonomy that have been influential in bioethics and

[1] See the papers in John Christman, *The Inner Citadel: Essays on Individual Autonomy*, Oxford University Press, 1989; Ruth Faden and Thomas Beauchamp, *A History and Theory of Informed Consent*, Oxford University Press, 1986, 8; Douglas Husak, 'Liberal Neutrality, Autonomy and Drug Prohibition', *Philosophy and Public Affairs*, 29, 2000, 43–80, who asserts bluntly that 'choice is valuable and worthy of respect when it is autonomous but not otherwise', p. 62.

beyond shows that weight, perhaps excessive weight, is sometimes placed on quite minimal and even implausible conceptions of individual autonomy.

2.2 INDIVIDUAL AUTONOMY IN A NATURALISTIC SETTING: MILL

Although a notion of autonomy can be traced back to antiquity, the conception of *individual autonomy* is less venerable. In antiquity the term *autonomy* was used to refer not to individuals, but to cities that made their own laws. An autonomous city was to be contrasted with a colony, whose laws were given, or rather imposed, by its mother city. The term *autonomy* gained a renewed but still restricted and political use in the early modern period, and its tremendous resonance in contemporary philosophical and political thought supposedly derives from the central role Immanuel Kant gave it in his moral philosophy. Kant, as is well known, saw autonomy as fundamental to morality.

Contemporary writing on autonomy often claims that it is fundamental in the first instance to individual agents, rather than to morality,[2] and introduces the idea that autonomy is fundamental or important to morality as a second move, by arguing for the importance of protecting, respecting and fostering individual autonomy. Individual autonomy is generally depicted as a capacity or trait that individuals may have to a greater or lesser degree, which they will manifest by acting independently, in the right and appropriate way.

The central difficulty for many current accounts of individual autonomy is that its proponents generally also take a naturalistic view of human action. John Stuart Mill first attempted the difficult philosophical task of incorporating an account of individual autonomy into a naturalistic account of action. In many writings, and in particular in *On Liberty*,[3] he sets out the earliest, the most famous

[2] See chapter 4.

[3] John Stuart Mill, *On Liberty* (1863), in *Utilitarianism, On Liberty and other Essays*, ed. Mary Warnock, Fontana, 1962.

and the most imitated attempt to place an account of individual autonomy within a naturalistic setting. Contemporary admiration for individual or personal autonomy still owes, I believe, far more to Mill than to Kant: although many of its admirers crave and claim Kantian credentials, they mostly seek an account of individual autonomy that fits within a naturalistic account of human action.

Naturalists view human action as caused by natural states and events, in particular by desires and beliefs. If they seek to fit an account of autonomy into a naturalistic account of action, they have to show which sorts of antecedents make an action independent in the relevant way, in short (and somewhat paradoxically) which sorts of dependence make an action autonomous. Yet once a naturalistic account of action is accepted, any view that choices or action are independent in ways that are morally significant merely because they have a distinctive sort of origin or cause is problematic. Why, we may ask, should any naturalist think that the fact that action arises via one rather than another type of causal pathway invariably make it more independent, or more valuable than it would have been if it had arisen by another causal pathway? We can hardly do better in considering this question than to turn briefly to John Stuart Mill, whose writings contain, in my view, the most profound attempt to set autonomy within a naturalistic account of human action.[4]

Mill hardly ever uses the word *autonomy*. I have found only a passing reference to the autonomy *of states* and none to the autonomy *of individuals* in his writings. I suspect that for Mill the term *autonomy* was too closely allied to Kant's non-naturalistic views of freedom and reason, which he emphatically rejects. Although Mill does not appropriate the term *autonomy*, or try to make it serve new purposes, many of his recent commentators use it very freely in interpreting his thought.[5]

[4] See John Skorupski, *John Stuart Mill*, Routledge, 1989, for a penetrating account of Mill's approach to this issue.

[5] For example, John Skorupski writes 'Mill in whose philosophy naturalism and . . . rational autonomy are the two deepest convictions is committed to the assumption that they are indeed reconcilable', *John Stuart Mill*, 43.

Mill himself argues that 'Civil or Social Liberty'[6] is the only way to secure the development and flourishing of '*persons of individuality and character*', that is of persons who have (a particular version of) what is now usually called personal or individual autonomy. Such persons, Mill argues, can flourish only if they enjoy protection not only against the 'tyranny of the magistrate' – the tyranny of despots and dictators – but also against the less obvious 'tyranny of the majority' and 'tyranny of society'.[7] Mill therefore opposes

The tendency of society to impose, by other means than civil penalties, its own ideas and practices as rules of conduct on those who dissent from them; to fetter the development and, if possible prevent the formation, of any individuality not in harmony with its ways.[8]

Mill's version of autonomy within a naturalistic setting sees individuals not merely as choosing to implement whatever desires they happen to have at a given moment, but as taking charge of those desires, as reflecting on and selecting among them in distinctive ways:

A person whose desires and impulses are his own – are the expression of his own nature, as it has been developed and modified by his own culture – is said to have a character. One whose desires and impulses are not his own, has no character, no more than a steam engine has a character.[9]

For Mill mere choice, mere action on whatever desires one happens to have at a given time, does not manifest individuality or character. Character and individuality require persons to 'own' or identify with certain desires, to cultivate certain feelings and impulses rather than others, thereby becoming well-developed human beings.[10] In a naturalistic theory of action these processes must, of course, arise from naturally occurring features of agents.

Mill uses this account of the formation of character as the basis for important normative claims. He holds that persons of marked

[6] Mill, *On Liberty*, 126. [7] Mill, *On Liberty*, 129–30. [8] Mill, *On Liberty*, 130

[9] Mill, *On Liberty*, 189; cf. Skorupski, *John Stuart Mill*, 348; also p. 43.

[10] Mill, *On Liberty*, 193.

character or individuality contribute pre-eminently to the well-being of humankind, making 'the free development of individuality . . . one of the leading essentials of well-being',[11] construed broadly as 'grounded in the permanent interests of man as a progressive being'.[12] By adopting this distinctive view of the sources of Utility or well-being, Mill can argue plausibly that individuality, and the liberty that protects it, are essential for utility; and more specifically that liberty is necessary for each to cultivate his or her own individuality and character and so to contribute both to individual and to social well-being. Hence Mill's firm and famous view that 'there is a limit to the legitimate interference of collective opinion with individual independence',[13] and that liberty is 'the only unfailing and permanent source of improvement'.[14] His conclusion is that, contrary to initial appearances, Utilitarian reasoning will show that a very extensive respect for individual liberty is morally required, indeed that the 'sole end for which mankind are warranted, individually or collectively, in interfering with the liberty of action of any of their number, is self-protection'.[15]

Mill's reasons for demanding an extensive liberty for individuals are therefore far stronger than those that a simpler utilitarian could give, who thinks that the satisfaction of all and any desires is valuable, who just has to hope that overall some of the nastier desires get outweighed, and who must accept that the value of liberty will vary with the actual configuration of desires. A simpler utilitarian has to allow satisfaction of sadistic and selfish desires to count at face value towards well-being, and has to take at their reduced face value desires that are blunted by or adapted to vile circumstances;[16]

[11] Mill, *On Liberty*, 185. [12] Mill, *On Liberty*, 136.

[13] John Skorupski renders Mill's view in the words: 'Autonomy – the freedom to make one's own private decisions in one's private domain – is in its own right a categorical human end, one of the essentials of a worthwhile human life', *John Stuart Mill*, 21.

[14] Mill, *On Liberty*, 200. [15] Mill, *On Liberty*, 135.

[16] For discussions of adaptive preferences and their implications for utilitarian thinking see Jon Elster, 'Sour Grapes – Utilitarianism and the Genesis of Wants', in Amartya Sen and Bernard Williams, eds., *Utilitarianism and Beyond*, Cambridge University Press, 1982, 219–38, reprinted in Christman, *The Inner Citadel*, 170–88; Jon Elster, *Sour Grapes: Studies in the Subversion of Rationality*, Cambridge University Press, 1983; and

a simpler utilitarian may well argue for extensive paternalism, including extensive social and legal controls, and will assign weight to widely held desires that those with unpopular views and tastes be forbidden to express them or to act on them. Unless simpler utilitarians hold a strong (and perhaps implausible) view that altruistic or sociable desires *invariably* outweigh hostile or competitive desires, and so find a different route to Mill's conclusions, they are likely reach pretty luke-warm views of the value of liberty.

Mill's normative arguments are impressive. As we warm to his conclusions it is easy to forget that they can be established only if he can show how the picture of self-mastery and of developing one's own character and individuality – his version of individual autonomy – fits within the naturalistic account of human action that he propounds. There are those who think that he succeeds in this task, that he is 'able to retain the deep liberal insight that freedom is rational autonomy, but without Kantian transcendentalism'.[17] Others doubt whether the enterprise works, or could work.

Similar approaches to individual or personal autonomy have been proposed and adopted by many late-twentieth-century writers. Like Mill, they think a naturalistic account of human action can be given, and specifically that all action has beliefs and desires among its causes. And like Mill they think that some action is nevertheless distinctively autonomous because it arises from desires of a distinctive sort. Contemporary accounts of individual autonomy are varied, complex and intriguing. Many writers have argued that autonomous choices are products of desires that the agent has controlled, or moderated, or endorsed, using other desires and beliefs. Autonomous choices are distinguished from mere choices by the fact that they follow from and reflect a greater degree of self-knowledge, or of self-control, or of capacities to review, revise and endorse other desires. Various versions of these thoughts about what makes autonomous choosing distinctively independent can be found in much contemporary discussion of human

Martha Nussbaum, *Women and Human Development: The Capabilities Approach*, Cambridge University Press, 2000.

[17] Skorupski, *John Stuart Mill*, 254; see also p. 43.

freedom and of ethics; they are repeated and elaborated in writing on bioethics.[18]

Some influential current versions of individual autonomy identify it specifically with a formal relationship between desires: choosing is autonomous when the first-order desires that it satisfies are backed by second-order desires (desires to have the first-order desires).[19] Others appeal to the idea of desires that survive reflective scrutiny; or of desires that have been fully identified with and integrated into the agent's character. Of course, within a naturalistic setting each of these neo-Millian processes of second-order endorsement, or reflective scrutiny, or of identification or endorsement that genuinely expresses the self or individuality of a person, must itself be seen as the outcome of a natural process. It is not at all obvious why these more elaborate causal processes secure a form of independence that merely spontaneous, unreflecting choosing lacks, or why the choices to which they lead should always (or generally) be more valuable.

2.3 THE TRIUMPH OF AUTONOMY

Despite these unresolved difficulties in accounts of individual or personal autonomy, appeals to individual autonomy have acquired tremendous resonance in bioethics. Autonomy gained its prominence in the field partly through the wide influence of Thomas Beauchamp and James Childress' *Principles of Biomedical Ethics* that has gone through numerous editions and been widely used as a textbook.[20] In it they argue that four principles are fundamental to bioethics: the principles of beneficence, of non-maleficence, of

[18] Various versions of autonomy within a naturalistic account of action are discussed in John Christman, 'Constructing the Inner Citadel: Recent Work on the Concept of Autonomy', *Ethics*, 99, 1988, 109–24; a number of influential papers are reprinted in Christman, *The Inner Citadel*; see also Faden and Beauchamp, *A History and Theory of Informed Consent*.

[19] Harry Frankfort 'Freedom of the Will and the Concept of a Person', *Journal of Philosophy*, 68, 1971, 5–20; reprinted in Christman, *The Inner Citadel*, 63–90.

[20] Thomas L. Beauchamp and James F. Childress, *Principles of Biomedical Ethics*, Oxford University Press, 1984; 4th edn., 1994.

autonomy and of justice. There has, of course, been plenty of debate about these four principles and their interpretation. Does not beneficence comprise non-maleficence, making one of the four principles redundant? Alternatively, if non-maleficence – the Hippocratic principle – is interpreted as an independent principle, how will it constrain the other principles, and in particular beneficence? Is it always wrong to do harm for the sake of a greater good? How are the principles to be justified? How are conflicts between them to be handled? Is not a principle of justice a different type of principle from the others, relevant to medical policies and systems rather than to clinical decisions and doctor–patient relationships? How is the move from principle to action in particular cases to be made?

Within a fairly short time the principle of autonomy had gained great support, and also attracted a lot of criticism. In the 1994 edition of their textbook, Beauchamp and Childress recognised that the principle of autonomy was sometimes being seen as *more* important in medical ethics even than the principle of beneficence:

> Whether respect for the autonomy of patients should have priority over professional beneficence has become a central problem in biomedical ethics. For proponents of autonomy rights for patients, the physician's obligations to the patient of disclosure, seeking consent, confidentiality and privacy are established primarily (and perhaps exclusively) by the principle of respect for autonomy. Others by contrast ground such obligations on the professional's obligatory beneficence. The physician's primary obligation is to act for the patient's medical benefit, not to promote autonomous decision-making. However, autonomy rights have become so influential that it is today difficult to find clear affirmations of traditional models of medical beneficence.[21]

To some commentators it seemed likely that this great emphasis on individual autonomy would be temporary, and would recede once

[21] Beauchamp and Childress, *Principles*, 272; for a historical account of the centrality of beneficence (and non-maleficence) in earlier medical ethics and of the rise and rise of autonomy at the expense of beneficence in the late twentieth century, see also Faden and Beauchamp, *A History and Theory of Informed Consent*, chapter 3.

paternalistic medical practices had been reformed,[22] and some have regretted the extent to which autonomy has become central, especially in work in bioethics in the USA. For example, Daniel Callahan wrote bluntly on retiring as Director of the Hastings Center.[23] 'Nothing has exasperated me so much as the deference given in bioethics to the principle of autonomy.'[24] His voice and many others have had little effect. Rather, as a more recent commentator suggests:

> For better or for worse (and in opposition to Beauchamp and Childress's model) autonomy has emerged as the most powerful principle in American bioethics, the basis of much theory and much regulation, and has become the 'default' principle . . . Indisputably . . . patient autonomy has become the most powerful principle in ethical decision making in American medicine.[25]

I do not think that this is the case to anything like the same extent in debates on this side of the Atlantic. Yet beyond the USA as well, many changes have been made in medical practice because they are thought to contribute to patient autonomy. They range from the huge reduction in compulsory detention of the mentally disturbed to the increased emphasis of formalised consent procedures not only for research participation but also for treatment; from proselytising for greater patient choice to demands that any advice or counselling be 'non-directive'. As we shall see, appeals

[22] Robert M. Veatch, 'Autonomy's Temporary Triumph', *Hastings Center Report*, 14, 1984, 38–40.

[23] The Hastings Center is one of the pioneering institutions for the study of bioethics. It is a 'non-profit, non-partisan organisation that carries out educational and research programs on ethical issues in medicine, the life sciences and the professions'; it was founded in 1969 and is located in Hastings-on-Hudson, New York; See institutional bibliography.

[24] Daniel Callahan, 'Can the Moral Commons Survive Autonomy?', *Hastings Center Report*, 26, 1996, 41–2. See also his 'Autonomy: A Moral Good not a Moral Obsession', *Hastings Center Report*, 14, 1984, 40–2.

[25] Paul Root Wolpe, 'The Triumph of Autonomy in American Bioethics': A Sociological View, in Raymond DeVries and Janardan Subedi, eds., *Bioethics and Society: Constructing the Ethical Enterprise*, Prentice-Hall, 1998, 38–59, p. 43.

to individual autonomy have gained special resonance in some discussions of new reproductive and genetic technologies.

2.4 THE TRIUMPH OF INFORMED CONSENT

Yet what does the supposed triumph of autonomy in medical ethics amount to? No doubt many of those who emphasise its importance have in mind some background naturalistic account of human action. No doubt also many have in mind some version of the Millian arguments for the importance of individuality and character, or of neo-Millian arguments for second-order endorsement or reflective choosing (generally, it must be said, they keep these thoughts rather far in the background). These rather than mere, sheer choice are what make individual autonomy *seem* important and appealing. But the practices that are proposed for securing or respecting autonomy in medical contexts are in fact generally no more than informed consent requirements.

By insisting on the importance of informed consent we *make* it *possible* for individuals to choose autonomously, however that it is to be construed. But we in no way guarantee or require that they do so. Those who insist on the importance of informed consent in medical practice typically say nothing about individuality or character, about self-mastery, or reflective endorsement, or self-control, or rational reflection, or second-order desires, or about any of the other specific ways in which autonomous choices supposedly are to be distinguished from other, mere choices.

In short, the focus of bioethical discussions of autonomy is not on patient autonomy or individual autonomy of any distinctive sort. What is rather grandly called 'patient autonomy' often amounts simply to a right to choose or refuse treatments on offer, and the corresponding obligations of practitioners not to proceed without patients' consent. Of course, some patients may use this liberty to accept or refuse treatment with a high degree of reflection and individuality, hence (on some accounts) with a high degree of individual or personal autonomy. But this need not generally be the case.

Requirements for informed consent are relevant to specifically autonomous choice only because they are relevant to choice of all sorts. What passes for patient autonomy in medical practice is operationalised by practices of informed consent: the much-discussed triumph of autonomy is mostly a triumph of informed consent requirements.

This minimalist interpretation of individual or personal autonomy in medical ethics in fact fits rather well with medical practice. When we are ill or injured we often find it hard to achieve any demanding version of individual autonomy. We are all too aware of our need and ignorance, and specifically that we need help from others whose expertise, control of resources and willingness to assist is not guaranteed. A person who is ill or injured is highly vulnerable to others, and highly dependent on their action and competence. Robust conceptions of autonomy may seem a burden and even unachievable for patients; mere choosing may be hard enough. And, in fact, the choices that patients are required to make are typically quite limited. It is not as if doctors offer patients a smorgasbord of possible treatments and interventions, a variegated menu of care and cure. Typically a diagnosis is followed with an indication of prognosis and suggestions for treatment to be undertaken. Patients are typically asked to choose from a smallish menu – often a menu of one item – that others have composed and described in simplified terms. This may suit us well when ill, but it is a far cry from any demanding exercise of individual autonomy.

It is probably a considerable relief to many patients that they are not asked to muster much in the way of individual autonomy. When we are ill or injured we often lack the skills or energy for demanding cognitive tasks. Our highest priority is to get help from others and in particular from others with relevant skills and knowledge. The traditional construction of doctor–patient relations as relations of trust, as quasi-personal, as guided by professional concern for the patient's best interests makes sense to many patients because (if achievable) it would secure what they most need. The point and the context of the older, trust-centred model of doctor–patient relationships are not at all obscure.

However, at a time at which the real relations between doctors and patients are no longer personal relationships, nor even one-to-one relationships, but rather relationships between patients and complex organisations staffed by many professionals, the older personal, trust-based model of doctor–patient relationships seems increasingly obsolete. Contemporary relations between professionals and patients are constrained, formalised and regulated in many ways, and may erode patients' reasons for trusting. The very requirements to record and file medical information, for example, while intended to control information and protect patients, can inhibit doctors' abilities to communicate freely. Doctors, like many other professionals, find themselves pressed to be accountable rather than to be communicative, to conform to regulations rather than to enter relations of trust. As layers of regulation and control are added with the aim of protecting dependent, ignorant and vulnerable patients, as professionals are disciplined by multiple systems of accountability backed by threats of litigation on grounds of professional negligence in case of failure to meet these requirements, relations between patients and professionals are inevitably reshaped. Much is demanded of informed consent requirements if they are to substitute for forms of trust that are no longer achievable (or perhaps were never widely achieved, and still less widely warranted), and safeguard the interests of patients who find strangers at their bedsides.

Looking back, we can note that many of the steps taken to protect patients from abuse of trust were not originally intended as ways of improving respect for patient autonomy. Informed consent requirements themselves are important and of benefit quite apart from considerations of patient autonomy. So are requirements for professional education and certification, entitlements to second opinions, requirements for professional indemnity insurance and well-designed complaints procedures. These are all of them measures that can benefit patients, that can protect professionals against litigation and that can even contribute to trust in the new contexts of hospital-based medicine.

2.5 IMPAIRED CAPACITIES TO CONSENT

Informed consent requirements have, of course, a long history and wide currency beyond bioethics, and more generally beyond medicine. They are widely seen as supplying at least part of the moral basis of countless institutions and practices: for government and for contracts, for shopping and for trade, for marriage and for employment. It may seem obvious that this well-understood constraint should be incorporated into medical practice, and obvious why discussion of informed consent has become the bread and butter of bioethical debate.

However, because of the vulnerability and weakness of patients, even informed consent procedures (let alone more robust conceptions of individual autonomy) may be less apposite and less realistic in medical practice than in almost any other area of life. Discussions of the importance of informed consent in other areas of life generally presuppose that we are dealing with people who are (to use Mill's famous phrase) 'in the maturity of their faculties'.[26] In medical practice this assumption fails in very many cases. Consent cannot be given by children, at least not by younger children; it cannot be given by patients who are seriously deranged (however temporarily); it cannot be given by those with learning disabilities (or at most in highly simplified form); it cannot be given by patients with dementia; it cannot be given by patients who are traumatised or unconscious; it often cannot be given in medical emergencies. One might add that most of us (even when in the maturity of our faculties), find it hard to express our individuality or independence, and even to muster the presence of mind needed for giving informed consent, when we are ill. Given that informed consent is problematic for so many patients, it can hardly be necessary for medical treatment.

No wonder then that a great deal of effort in bioethics has gone into considering the proper treatment of those who are not, or not fully, competent to consent. A truly vast literature, both professional and philosophical, has explored the *hard cases for informed*

[26] Mill, *On Liberty*, 135.

consent requirements. The questions are endless and perplexing. What are the limits of relying on the proxy consent given by parents to the treatment of children? When may courts intervene on behalf of a child whose parents are inclined to make decisions that professionals do not think in the child's best interests? At what age should we seek the consent of older children, and for which sorts of treatment? May parents determine whether teenage children – so-called *mature minors* – should be allowed contraception? When should parents be allowed to decide whether one child should undergo tests or investigation for the benefit of a sibling? Is parental consent sufficient to allow blood tests, or genetic tests, or 'donating' bone marrow for the benefit of a brother or sister? Could parental consent suffice for the 'donation' of a kidney? Is non-therapeutic research on children ever acceptable?[27]

Equally difficult questions arise for many adults. Under what conditions may adults be treated in defiance of their currently expressed wishes (for example, during mental disturbance) or at variance with their previously expressed wishes (for example, those expressed in 'living wills' or prior declarations)? To what extent should adults with learning difficulties be asked whether they agree to their treatment, and what should be done when they cannot do so, or when they refuse needed treatment? What is the standing of next-of-kin, or guardians or friends when illness or incapacity prevents an adult patient from consenting?

Beyond the arena of decisions on medical treatment a whole range of further questions can arise in making reproductive decisions and seeking genetic information. When may parents, or guardians, or professionals give consent on behalf of adults who are not competent to consent themselves to interventions such as contraception, abortion or sterilisation? When may parents, or guardians, or professionals seek genetic information about those who are not competent to consent to providing that information?

And beyond all these clinical questions there are difficulties that arise when consent is sought from research subjects. Do volunteers

[27] See Thomas H. Murray, *The Worth of a Child*, University of California Press, 1996, chapter 4.

in clinical trials give sufficiently informed consent when an experimental design requires them not to be informed whether the pill they are to swallow is a placebo or potentially effective? Should vulnerable people – prisoners, the mentally disturbed and the frail elderly – be allowed to become research subjects, or do these vulnerabilities invariably compromise the quality of consent? How are clinical trials to be undertaken for drugs designed for conditions that impair capacities to consent (including Alzheimer's and Parkinson's)? What standards should pharmaceutical companies from richer countries maintain in seeking informed consent from research subjects in poorer countries? What consent requirements, if any, are needed for secondary analyses of medical data collected without explicit consent in the course of previous treatment?[28]

These and many similar questions have been central to day-to-day discussions of medical ethics. They are essentially thought of as questions about the *borderline cases*. Those who discuss them assume that in the *normal case* of the adult in 'the maturity of his faculties' informed consent procedures do indeed protect patients, and allow for at least a minimal form of individual autonomy when patients can muster it. The principal difficulty, supposedly, is to devise procedures that give equivalent protection to those not in the maturity of their faculties. This limited view of ways in which informed consent procedures can be problematic seems to me to overlook larger difficulties.

2.6 CONSENT AND OPACITY

Informed consent procedures can, I think, take only a more limited role in protecting patient autonomy, and in protecting patient well-being, than this concentration on hard cases suggests. This is not because informed consent procedures fail to guarantee robust or distinctive forms of individual autonomy in the choice of medical treatment. The deficiency is more basic and is common to all

[28] House of Lords, Select Committee on Science and Technology IIa, *Report on Human Genetic Databases: Challenges and Opportunities*, HL 57, 2001; written evidence, HL 115, 2000; see institutional bibliography.

reliance on informed consent procedures, including those that are incorporated in all the little transactions of daily life such as shopping and ordering a meal, using public transport and filling the car with petrol. Informed consent requirements protect routine, habitual, day-to-day choosing and transactions as much as they protect choosing that shows more robust forms of individual autonomy. Consumer choice is protected and structured by informed consent requirements, although many consumer choices are not robustly or distinctively autonomous. I may be a competent customer even if I am a slave to fashion. Equally, when patients choose deferentially and unreflectively, so by many standards non-autonomously, it will still be true that informed consent requirements protect that limited exercise of choice.

The problem is rather that, whether choosing is routine or robustly independent, there are systematic limitations to the degree of justification that informed consent procedures can offer. These arise because informed consent is always given to one or another *description* of a *proposal* for treatment. Consent is a *propositional attitude*: it has as its object not a procedure or treatment, but rather one or another proposition containing a description of the intended procedure or treatment.

In consequence consent, like other cognitive attitudes that take propositions as their object (such as knowing, believing, desiring or trusting), is *opaque*.[29] In consenting to a proposition I may see no further than the specific descriptions that it contains. I may not see myself as consenting to other equivalent or closely related propositions, or to propositions that are entailed by those to which I consent. In addition I well not be aware of, *a fortiori* not consent to, the standard and foreseeable consequences of that to which I consent. In the first case I might consent to a medical procedure described in euphemistic and unthreatening ways, yet not see myself as consenting to another more forthright and equivalent description of that very treatment. In the second case I might consent to chemotherapy, and yet when as a result I feel desperately ill

[29] W. V. O. Quine, 'Two Dogmas of Empiricism' in Quine, *From a Logical Point of View: 9 Logico-Philosophical Essays*, 2nd edn., Harper & Row, 1963, 20–46.

and weak may truthfully claim that I never consented to anything that would have *this* effect – even if these very effects were carefully described as among the normal effects of the treatment. I may consent to transplant surgery or to an amputation, and be given extensive information, yet later feel that I never consented to what has actually been done to me. I put these points about *referential opacity* and *failure to grasp the consequences* briefly; but their implications are profound. They show that informed consent can be quite superficial, fastening on the actual phrases and descriptions used, and need not take on board much that is closely connected to, even entailed by, those phrases and descriptions.

In addition to these theoretical limits on the reach of all consent, there are practical limits to all informed consent requirements, even where patients are uncontroversially in the maturity of their faculties. These may arise because medical information is complex and professional time scarce and expensive, or because most of us when feeling ill are less than perfect at eliciting and assimilating information that can be complex and upsetting.

Although the phrase 'fully informed consent' is frequently and approvingly mouthed, full disclosure of information is neither definable nor achievable; and even if it could be provided, there is little chance of its comprehensive assimilation. At best we may hope that consent given by patients in the maturity of their faculties, although not based on full information, will be based on a reasonably honest and not radically or materially incomplete accounts of intended treatment, and that patients understand these accounts and their more central implications and consequences to a reasonable degree. Even this level of informed consent demands a lot both of patients and of professionals, although it does not demand any striking form of individual autonomy from either.

2.7 THE CONSUMER VIEW OF AUTONOMY

Given these intrinsic and practical limitations of informed consent procedures, how much can they justify? In particular is informed consent, realistically understood, either necessary or sufficient justification for action, and in particular for medical intervention?

We have seen that the classical arguments for liberty that we inherit from John Stuart Mill's writings never view consent, nor therefore informed consent, as necessary for all justifiable action. Mill allows for restrictions of liberty where its exercise will harm others. The predominant view in medicine and in contemporary bioethics has been similar. The general requirement of consent to medical treatment is always hedged with provisos permitting treatment without consent where refraining from treating would harm others. Consent is at most seen as *generally*, rather than *invariably*, necessary for justifiable medical treatment.

Classical examples where mandatory or forced medical treatment has been viewed as permissible in order to prevent harm to others include the use of detention and quarantine for public health reasons and unconsented-to treatment of those likely to harm others (for example, during psychotic episodes). More recent (and controversial) examples might include mandatory HIV tests for those doing certain sorts of work and mandatory caesareans for women whose refusal endangers their infant.

Consent requirements are particularly problematic in public health, environmental and food safety decisions. Here policies of limiting individual autonomy only to prevent harm to others offer inadequate guidance. For example, environmental policies and public health measures aimed at ensuring safe drinking water or food hygiene usually have costs as well as benefits. It is likely that every possible policy will harm or inconvenience some people, and very likely that no policy will receive unanimous consent. Concern for individual autonomy, even for individual informed consent, cannot play a large part in public health or environmental decisions where this is the case.[30]

There are similar difficulties in applying informed consent standards in areas of medicine where individual decisions have public

[30] Some think that processes of democratic legitimation can supply the missing justification: see chapter 8.

health implications. To take a topical example: if a few parents refuse consent to their children's vaccination against a dangerous disease they risk little harm to their own or to others' children. A small unvaccinated minority can 'free ride' and remain healthy amid a vaccinated majority. But as more and more parents take this view, free riding fails and harm is risked to one's own and to others' children. Yet it is impossible to demonstrate that an individual parent who refuses vaccination thereby harms their own or others' children. If the issue is looked at solely from the perspective of the individual parent, grounds for overriding their refusal and requiring vaccination may seem weak, whatever the public health costs.

Analogous problems constantly arise in environmental decisions: there is often little harm in any individual acting in an environmentally unfriendly way, making it difficult to see whether a Millian line of thought could ever justify prohibiting such action. An individual who uses a tiny amount of some non-renewable resource, or who creates a small amount of pollution or makes a small change in a landscape of outstanding beauty does little harm: yet the unintended consequences of such action by many individuals may create large and long-term harm. As the probable cumulative damage mounts, so does the case for restricting liberty. In such cases it is once again hard to demonstrate that a particular act will harm, hence hard to show that or when liberty may be overridden by appeal solely to the avoidance of harm. Reasons for appealing to *multiple* principles in bioethics, if not necessarily to the famous four (non-maleficence, beneficence, autonomy and justice) are strengthened by these standard instances of difficulties that arise in bioethical arguments that appeal only to individual autonomy and the avoidance of harm. If the literatures on rational choice have shown anything, they have shown the disjunction between individual and collective choice is deeper than it once seemed. Nevertheless individual autonomy has continued to ride high in bioethical debate.

In stressing the importance of individual autonomy, Beauchamp and Childress did not intend that it become the overriding concern or the default principle in bioethics. Nor, as we have seen does individual autonomy provide a plausible basis for the full range of issues that arise even in medical ethics. How and why has the prevailing tendency to treat autonomy as the key principle of bioethics emerged?

One plausible source of the 'triumph' of autonomy, indeed the triumph of the minimal conception of individual autonomy that identifies it with informed consent requirements, may be that there is one presently prized domain of life in which informed consent requirements are often seen not only as *necessary* but also as *sufficient* for ethical justification. This is the area of consumer choice. If we abstract from all the fundamental ethical principles that structure the institutional background of consumer choice – such as those that shape and justify economic and legal institutions – we may very well think that informed consent is both the necessary and the sufficient justification of all economic transactions in the everyday life of a consumer society. If we extend the consumerist vision of informed consent into medicine, and beyond medicine into science and biotechnology, the justifications provided by informed consent may come to seem not only necessary but also sufficient in those domains as well. As one sociologist of medicine puts the point:

> In a world where medicine has become a good to be consumed, and where patients are customers to be wooed, informed consent becomes the disclosure of the contents on the back of the box. Informed consent involves discussion of the nature of a procedure, its risks and benefits, and alternative treatments, and is enacted through the modern ritual of free assent, the signing of a contract.[31]

Once medical decisions are conceived as falling within the domain of consumer choice, we are immediately distanced from the more complex conceptions of individuality that concerned Mill, and that many recent writers on individual autonomy have also found attractive. A reduced vision of individual autonomy is identified with the informed consent of those involved in a given transaction. Within this consumerist, quasi-libertarian framework the tasks of

[31] Wolpe, 'The Triumph of Autonomy in American Bioethics', 49.

medical ethics are concentrated in one idea: informed consent is seen as offering *necessary and sufficient ethical justification.*

Of course, even the consumerist vision has its complexities. Capacities to consent may fail; it can be hard to identify all parties to a transaction; it is unclear how to view unconsented-to costs imposed on those not directly involved in a transaction. Even ardent champions of consumerism sometimes doubt whether it provides a basis for professional practice, or whether the intrusion of advertising into professional life is acceptable. However, consumerist ideologies are worth thinking about because they propose a breathtaking simplification of ethical justification in and beyond medicine. The single requirement of respect for individual autonomy, under a strikingly weak interpretation that demands no more than respect for informed consent requirements, is seen as a complete basis for all ethical justification in medicine (perhaps also in science and biotechnology).

I believe that there are deep reasons for being sceptical of a consumerist view of justification in bioethics and beyond. A principal reason for thinking informed consent important in medical ethics is that it protects individual autonomy. Yet why should we think individual autonomy so important, and in particular why should we think it so fundamental for patients, whose primary need is for others' competent and willing help? Might the real importance of individual autonomy to medical ethics lie not in an illusory quest for patient autonomy, but in the great scope that certain new biotechnologies offer for the exercise of individual autonomy? Could the deeper reasons for prizing individual autonomy perhaps lie in its importance in making choices about human reproduction and human genetics?

CHAPTER THREE

'Reproductive autonomy' and new technologies

3.1 AUTONOMY AND TWENTIETH-CENTURY REPRODUCTION

In contemporary medical practice patient autonomy is often no more than a right to refuse treatment. This right is important. Insofar as patients are protected by informed consent procedures that are scrupulously used, they will be protected against coercive or deceptive medical treatment. However, by themselves informed consent procedures neither assume nor ensure that patients are autonomous in any more demanding sense. Patients who give or withhold informed consent may or may not have extensive capacities for self-determination, for reflective evaluation, or for independence. If they have any of these capacities, they may make little use of them in consenting or withholding consent from treatment.

This limited focus on informed consent, rather than on any more extensive conception of autonomy, serves reasonably well in medical ethics because it suits the real context of illness and injury. When we are patients we are not well placed to exercise any very demanding form of autonomy. As the very etymology of the word *patient* suggests, patients are likely to find robust forms of autonomy taxing if not impossible. Even informed consent procedures can be strenuous for the seriously ill or injured.

More demanding conceptions of individual autonomy may still be important in bioethics. Perhaps we need robust and distinctive capacities for autonomy if we are to make certain significant life choices of the types made available by new biotechnologies. Already reproductive and genetic technologies offer new choices, and wider ranges of choice are in prospect, at least in rich societies. Robust forms of individual autonomy may be more significant in the use of these biotechnologies than it is in the lives of patients.

One way of telling the history of reproductive medicine in the twentieth century would be to see it as a story of progress, and in particular of progress towards individual human autonomy. The terminology has of course varied: we recognise different contexts and commitments behind phrases such as *reproductive choice*, *planned parenthood*, *right to choose*, and latterly *reproductive freedom* and *reproductive autonomy*, *procreative freedom* and *procreative autonomy*. The changing vocabulary reflects changes in reproductive technologies and opportunities, as well as different views about their ethically acceptable use.

The importance of reproductive freedom in the widest sense of the term in all human lives is underlined by the fact that during the twentieth century it was at times hideously violated. In the early twentieth century there was forced sterilisation in many countries (often with eugenic aims); in mid-century there was forced childbearing under the Nazis; later there was forced abortion in China. Through most of the century and in many societies an assumption that forced sex in marriage does not count as rape allowed for further reproductive coercion. In some societies forced marriage remained common, and again created a context for reproductive coercion. Reproductive coercion is at least as bad as any other form of coercion, and worse than many. Even if reproductive freedom were only a matter of ending and forbidding these practices, it would be ethically significant, and would secure a degree of choice whether or not to have a (further) child. Protecting this choice is of great importance, whether or not those who choose are capable of or deploy any very demanding or distinctive form of autonomy.

However, *reproductive autonomy*, as the term is now generally understood, comprises far more than freedom from coercion in reproduction: the idea has been extended to reflect the expansion of possibilities for individual self-determination and independence offered by new reproductive technologies that began to be used in the last decades of the twentieth century.

At the start of the twentieth century many fertile women were either celibate or sexually constrained (whether within or outside marriage) or could expect pregnancies that might cost them ill health, or even their lives, and were likely to bear more children than they wanted or were able to care for. The physical demands of pregnancy, the long-term demands of motherhood and the economic obligations of parents all constrained and shaped lives and life possibilities for most people in profound ways: these realities were serious barriers to individual autonomy *however conceived* in many aspects of life. By the end of the century fertility control in richer societies was extensive, and could often be exercised by women; maternal and infant mortality were hugely reduced and maternal reproductive health had improved. Similar changes are now taking place in many poorer societies. Although populations are still growing at a frighteningly fast rate in some parts of the world – powered by increased longevity as much as by numbers of new lives – children in many parts of the world are now more likely than in the past to be chosen, spaced and presumably wanted.

Partly because of these advances, the focus of reproductive ethics by the end of the twentieth century had broadened to cover the use of a wide range of new reproductive technologies (NRTs). As this happened, appeals to individual autonomy in discussions of reproduction were transformed. This transformation may show where and why autonomy, robustly conceived, is important in bioethics.

3.2 THE 'RIGHT TO CHOOSE': CONTRACEPTION

Appeals to a supposed 'right to choose' were common and plausible in early and mid-twentieth-century battles to reform social legislation that governed reproductive medicine. The central demand of many reformers was that individuals should be permitted to control their fertility: they needed access to contraception in order to achieve planned parenthood. It is easy to forget how hard-fought these battles were. Historical work on the development of contraceptives reminds us that even scientific research on human reproduction and contraception, let alone making effective contraceptive advice and technology widely available, was opposed by many who feared that once it was available women would achieve greater independence, and so greater individual autonomy. Often the fear was that this independence would include sexual independence, that contraception would remove one of the great sanctions of women's marital fidelity and pre-marital chastity. Partly for this reason, much pioneering work on the physiology of reproduction was done not by medical researchers but in departments of agriculture, and much of it was not publicly funded but rather depended on the funding of one independent body, the Rockefeller Foundation.[1]

In this context we can make good sense of ideas such as *rights to choose* and *reproductive freedom*. As contraception became legal and available, women and couples acquired greater control over their reproduction. These rights gave them a greater control of reproductive health and family commitments, and consequently a greater prospect of choosing the sorts of lives they wished to lead. Having these choices they might then lead lives that manifested greater individual autonomy than lives that they could otherwise have led. What they gained was perhaps not *reproductive autonomy*, but rather greater individual autonomy in many aspects of life, achieved by better control of the timing and amount of reproduction. The new reproductive technologies of the early and mid-twentieth century brought benefits and freedom to many, without restricting others' freedom to live their lives and to do likewise – or otherwise – in matters of reproduction.

3.3 THE 'RIGHT TO CHOOSE': ABORTION

Notoriously some methods of fertility limitation are not universally accepted. The use of IUDs and morning-after pills, and above all abortion, are notoriously viewed as unacceptable by a significant minority; a much smaller minority thinks all contraception wrong.

[1] Adele E. Clarke, *Disciplining Reproduction: Modernity, American Life Sciences and 'The Problem of Sex'*, University of California Press, 1998.

The often-unexamined and highly ambiguous slogans *right to life* and *right to choose* became the preferred bludgeons in furious debates about the legalisation of abortion. Appeals to the *right to choose* were extended to express not only the idea that reproduction is in very many respects an area of life in which persons have a right to make their own choices, but the thought that it is a domain in which nobody else has any right to determine what they shall do in any respect.[2] Those advocates of a *right to life* who oppose abortion do not challenge the thought that there are *rights to choose* in many matters, including many aspects of reproduction. They deny quite specifically that there can be a *right to choose abortion*. They think quite specifically that from the moment of conception there is a further person, whose *right to life* trumps any supposed *right to choose* of a pregnant woman that would allow her to destroy the embryo or foetus.

The attempt to extend arguments that had been so compelling for justifying access to contraception met their first great challenge in the abortion debates of the 1960s and 1970s. The early stages of these debates, particularly as they took place in the USA, are quite surprising. If we cast our minds back to the period before contraception became widely available, we are also casting them back to a period before the emergence of contemporary bioethics, and in particular to a period in which medical paternalism had not been widely questioned. It was not in that context redundant to argue that women and couples should have a *right to choose* whether to use contraception. There were many jurisdictions in which contraception was illegal, and even more in which its availability was at the discretion of doctors. In the battle against paternalism in reproductive medicine, appeals to *autonomy*, to a *right to choose* and to a *right to privacy* became closely linked.

[2] Even the most ardent advocates of reproductive autonomy do not claim that there should be *no* regulation of human reproduction. For reminders see Ruth Deech, 'Cloning and Public Policy', in Justine Burley, ed., *The Genetic Revolution and Human Rights*, Oxford University Press, 1999, 95–100, esp. p. 97; for controversial thoughts see John Harris, 'Clones, Genes and Human Rights', in Burley, *The Genetic Revolution*, 61–94. For discussion of the range of possibilities see the works cited in n. 31.

The battle to make contraception more widely available was particularly fierce in the USA. In 1965 the US Supreme Court struck down the State of Connecticut law assigning control of contraception to doctors, arguing that this violated the *right to privacy* of the married couple, which included a *right to choose contraception*,[3] so was no matter for legislators. Much more surprisingly – at least to the European ear – in 1973 the Supreme Court once again appealed to a supposed *right to privacy* in the even more famous case in which it struck down Texas legislation outlawing abortion. The majority of the Court concluded, 'that the right of personal privacy includes the abortion decision, but that this right is not unqualified and must be considered against important state interests in regulation'.[4]

The *right to privacy* has not always been part of the US constitutional tradition. It is not mentioned in the Constitution or the Bill of Rights. In giving judgement in *Roe* v. *Wade* the Supreme Court acknowledged this:

> The Constitution does not explicitly mention any right of privacy. In a line of decisions, however, going back perhaps as far *as Union Pacific R. Co.* v. *Botsford*, 141 US 250, 251 (1891), the Court has recognised that a right of personal privacy, or a guarantee of certain areas or zones of privacy, does exist under the Constitution.[5]

Particular weight was given in this judgement to a classical article on the *right to privacy*, whose leading author later became Mr Justice Brandeis. It begins:

> That the individual shall have full protection in person and in property is a principle as old as the common law; but it has been found necessary from time to time to define anew the exact nature and extent of such protection. Political, social, and economic changes entail the recognition of new rights, and the common law, in its eternal youth, grows to meet the demands of society. Thus, in very early times, the law gave a remedy only for physical interference with life and property, for trespasses *vi et armis*. Then the 'right to life' served only to protect the subject from battery in

[3] *Griswold* v. *Connecticut*, 381 US 479 (1965). [4] *Roe* v. *Wade*, 410 US 113 (1973).
[5] *Roe* v. *Wade*, Mr Justice Blackmun delivering judgement for the court, section VIII.

its various forms; liberty meant freedom from actual restraint; and the right to property secured to the individual his lands and his cattle. Later, there came a recognition of man's spiritual nature, of his feelings and his intellect. Gradually, the scope of these legal rights broadened, and now the right to life has come to mean the right to enjoy life – the right to be let alone; the right to liberty secures the exercise of extensive civil privileges; and the term 'property' has grown to comprise every form of possession–intangible as well as tangible.[6]

It is striking that Brandeis grounds *rights to choose* and *rights to privacy* in the *right to life*, in sharp contrast to those who now claim that in human reproduction any *right to choose* and any *right to privacy* are limited by the *right to life*! We can at least read this oddity as a warning that appeals to supposed rights picked out in substantival terms are ill-defined, and open to many different interpretations; I shall return to this point in chapter 4. Nevertheless, the legalisation of abortion has commonly been debated in these tangled terms both in and beyond the USA for thirty years. During this period the *right to choose* has been increasingly portrayed as much more than a right not to be forced in matters of reproduction, for instance as a *right to autonomy* or *right to privacy*, or as Brandeis also has it as a *right to be let alone*, and (as others have put it) as a *right to self determination*.[7]

Reproductive choice, variously understood, was increasingly seen as a vital component of individual or personal autonomy. Some writers have argued that reproductive choice is a profound form of self-expression, an exercise of individual autonomy that goes far beyond mere assent to or dissent from others' proposals. One of the most ardent advocates of *procreative autonomy* in this deeper sense has been Ronald Dworkin, who defines it as 'a right [of people] to control their own role in procreation unless the state has compelling reasons for denying them that control'.[8] He argues that:

[6] Samuel D. Warren and Louis D. Brandeis, 'The Right to Privacy', *Harvard Law Review*, IV, 1890, 194–219.

[7] Allen Buchanan, Dan W. Brock, Norman Daniels and Daniel Wikler, *From Chance to Choice: Genetics and Justice*, Cambridge University Press, 2000, 218.

[8] Ronald Dworkin, *Life's Dominion*, HarperCollins, 1983, 148.

> The right of procreative autonomy has an important place not only in the structure of the American Constitution but also in Western Political Thought more generally. The most important feature of that culture is a belief in individual human dignity: that people have the moral right – and the moral responsibility – to confront the most fundamental questions about the meaning and value of their own lives for themselves: answering to their own conscience and convictions . . . The principle of procreative autonomy, in a broad sense, is embedded in any genuinely democratic culture.[9]

In this way of thinking, reproductive or procreative autonomy is on a par with freedom of thought and conscience, and of equally profound importance. If our reproductive decisions express our deepest sense of who and what we are, then the way is seemingly open for arguing that reproductive autonomy must be more than a right to accept or refuse reproductive treatments and technologies others offer, and that it must include a wide right to self-determination and self-expression in reproductive matters. Reproductive freedom looked at in this way is often taken to include not only a right to choose abortion, but a right to choose among new reproductive technologies, and as showing that neither choice should be subject to state prohibition or regulation unless there are clear harms to be prevented.

I shall not try to summarise decades of furious debate, but I think that going back to some of its sources reveals one point that is worth pausing on for present purposes, which is that the weight of appeals to *rights to choose, to privacy* or *to autonomy* is quite different in debates about contraception, about abortion and about the use of certain new reproductive technologies.

In the arguments against state or professional restrictions on access to contraception, the persons involved are indeed primarily – usually exclusively – the couple concerned, and above all the woman whose health and life will be so closely affected by having a (further) child. But in the case of abortion and of use of technologies for assisted reproduction, the possibility of a third

[9] Dworkin, *Life's Dominion*, 166–7.

party – successively embryo, foetus and child – is also at stake. Those who appeal to a supposed right to life that begins with conception may indeed lack a sound argument: for it may well be the case, as the US Supreme Court also held, and as many others have argued,[10] that the claim that the newly conceived embryo is a person is speculative, and that it confuses potential with actual persons. The view that the fertilised egg has the full rights of a person is not the ancient teaching of the Christian Church (which traditionally differentiated early from late abortions), but rather a nineteenth-century innovation, paralleled by other pieces of restrictive and intrusive social legislation. Nevertheless, *by itself* no appeal to a wholly indeterminate *right to choose* can make the case for permitting abortion. No argument can convince without showing that there is no *right to life* in the embryo or foetus that trumps the supposed *right to choose* of the mother or parents.

3.4 THE 'RIGHT TO CHOOSE': ASSISTED REPRODUCTIVE TECHNOLOGIES

Rather than entering further into debates about the status and rights of the embryo and the (early) foetus I want now to turn to the role that appeals to *reproductive autonomy* have played in discussions of more recent reproductive technologies. In the last decades of the twentieth century these technologies opened the way to possibilities of self-determination and self-expression in reproduction that went far beyond the avoidance of unwanted children and the timing and spacing of wanted children. Correspondingly appeals to *reproductive autonomy* or *procreative autonomy* became more varied and ambitious, at least in rich societies. Attention shifted from the problem of controlling unwanted fertility (although the abortion debate lost none of its steam) to that of dealing with unwanted infertility.[11] Appeals to autonomy were invoked to support use of

[10] Cogently John Harris, *Clones, Genes and Immortality: Ethics and the Genetic Revolution*, Oxford University Press, 1998a, esp. pp. 47–53.

[11] Gillian R. Bentley and C.G. Nicholas Mascie Taylor, eds., *Infertility in the Modern World: Present and Future Prospects*, Cambridge University Press, 2000.

(and even guaranteed and subsidised access to) a wide variety of assisted reproductive technologies, ranging from hormone treatment to IVF, to the use of eggs, sperm and gestation provided by others, from post-menopausal pregnancy and *post mortem* paternity, to cloning and the production of so-called 'designer babies'. By the end of the century a range of discussions of reproductive choices incorporating genetic aims had also begun. I shall touch only lightly on this area of more speculative discussion.

Somewhere in this welter of appeals and arguments involving conceptions of reproductive autonomy a case may have been made for a more robust, yet plausible conception of human autonomy. But if these appeals and arguments tacitly rely on inflated or implausible conceptions of human autonomy they will not convince. Of course, that will not show that using any specific reproductive technology is wrong, but only that appeals to reproductive autonomy alone cannot establish any rights to use specific technologies.

At the time of *Griswold* v. *Connecticut* and *Roe* v. *Wade* arguments were mainly focused on avoiding unwanted reproduction: the problem was unwanted fertility (still a pressing problem in much of the world). As reproductive technologies that could overcome some of the problems of the infertile were developed, arguments first used in discussions of unwanted fertility and abortion were redirected in claims about the use of the new reproductive technologies. *Reproductive autonomy* was repeatedly cited as a reason not to restrict the use of a wide range of fertility treatments.

For example, John Harris argues for extensive reproductive autonomy in order to resolve issues 'concerning the possible use of foetal ovarian tissues and eggs, including . . . about cloning human eggs . . . [and about] pre-natal and pre-implantation screening'.[12] He also argues that 'it seems invidious to require that people who need assistance with procreation meet tests to which those who need no such assistance are not subjected'.[13] The only reason he

[12] John Harris, 'Rights and Reproductive Choice', in John Harris and Søren Holm, eds., *The Future of Human Reproduction: Ethics, Choice and Regulation*, Clarendon Press, 1998b, 5–37, pp. 5–6.

[13] Harris, 'Rights and Reproductive Choice', 7.

sees for forbidding the use of a technology would be the likelihood of harm. Using this criterion Harris suggests that there are no good reasons to forbid the use of a number of reproductive technologies that are presently not permitted in the UK. In particular, he sees no good reason to forbid the use of cadaveric eggs, including their use to replace a lost child, or the payment of egg 'donors', or the use of foetal eggs, or the use of posthumous sperm; no reasons to forbid post-menopausal fertility treatment, or genetic enhancement (choosing genetic characteristics of future children); no reasons to forbid splitting embryos and implanting twin embryos successively, or to forbid selecting embryos for their gender. Needless to say, Harris also endorses the use of those reproductive technologies that are at present permitted in the UK (subject to regulation), such as IVF, use of donated sperm and eggs, as well as pre-implantation diagnosis (PID) and foetal genetic screening in cases where serious genetic illness may be inherited. 'Arguably', Harris writes, 'even freedom to clone one's own genes might also be defended as a dimension of procreative autonomy'.[14] In short, he defends an extensive conception of *reproductive autonomy*, which he suggests 'might be legitimately interpreted to include the right to reproduce with the genes we choose and to which we have legitimate access, or to reproduce in ways that express our reproductive choices and our vision for the sorts of people we think it right to create'.[15] People who exercise these supposed rights are conceived of as doing far more than consenting to or refusing treatments proposed by professionals: they are envisaged as choosing among possible lives and possible forms of (family) life.

Harris is not alone in holding these views. For example, John Robertson embraces an even more extended view of reproductive freedom. He argues that 'procreative liberty be given presumptive priority in all conflicts, with the burden on opponents of any particular technique to show that harmful effects of its use justify

[14] Harris, 'Rights and Reproductive Choice', 34–5.
[15] Harris, 'Rights and Reproductive Choice', 34.

limiting procreative choice',[16] and sees a market in reproductive factors and services as desirable because it will allow 'the invisible hand of procreative preference' to do its supposedly benign work.[17]

It may seem that we are landing on the wilder shores of reproductive fantasy, that this is an appeal to mere, sheer choice with scant regard for anything else. That is not how Ronald Dworkin and John Harris see matters. Both argue that the sorts of choices that are at stake in human reproduction are not mere choices, but that they are peculiarly intimately bound up with our deepest individual nature, and that they are central to individual autonomy, robustly construed. Dworkin in particular draws an analogy between reproductive freedom and freedom of expression. The difference of outlook between 'pro-life' and 'pro-choice' positions, as he sees it, is essentially a difference between two views of the sanctity of life, hence a religious disagreement and so one that deserves protection under the First Amendment of the US constitution. Dworkin writes:

> The right to procreative autonomy follows from any competent interpretation of the due process clause and of the Supreme Court's past decisions applying it . . . The First Amendment prohibits government from establishing any religion, and it guarantees all citizens free exercise of their own religion . . . These provisions also guarantee the right of procreative autonomy.[18]

Harris concurs in drawing a close analogy between rights to religious freedom and to procreative freedom. He writes:

[16] John A. Robertson, *Children of Choice: Freedom and the New Reproductive Technologies*, Princeton University Press, 1994, 16.

[17] John A. Robertson, 'Embryos, Families and Procreative Liberty: The Legal Structures of the New Reproduction', *Southern California Law Review*, 59, 1986, 939–1041, pp. 1030. He concludes that 'a regime of private discretion in attempts to procreate, with minimal regulation, must prevail', p. 1040.

[18] Ronald Dworkin, *Freedom's Law*, Oxford University Press, 1996, 104–5; see also pp. 237–8. This particular argument is relativised to the US Constitution so cannot deliver ethical justification; the arguments from individual autonomy aim higher.

The point is that the sorts of freedom which freedom of religion guarantees, freedom to choose one's own way of life according to one's own most deeply held beliefs, are also at the heart of procreative choices.[19]

Procreative autonomy, these writers argue more generally, should be extensive and protected because procreation, like religion, is an area of life in which we express our most intimate and personal choices. Increasingly advocates of procreative autonomy also seek to read procreative autonomy back into Article 16 of the Universal Declaration of Human Rights, which proclaims a right to found a family, so suggesting a right to have children.[20]

3.5 REPRODUCTIVE CHOICE AND PARENTHOOD

This metamorphosis of the *right to choose* into a *right to procreative autonomy* seems to me deeply unconvincing. Reproduction indeed matters to people; it is indeed a part of life in which they express their deepest beliefs. But it does not follow that it is or should be seen primarily as a matter of self-expression, or that it should be protected as we protect self-expression. Reproduction is unlike both contraception and abortion, in that it aims to bring a third party – a child – into existence. The misfortune of the infertile is that this is not readily achieved: that is why they seek assistance with reproduction. Reproductive choice is therefore not best seen on the model of the exercise of a liberty right, such as a right to freedom of expression. Ideals of individual or personal autonomy, whether thought of as a matter of independence, of self-determination, or of self-expression are unpromising starting points for thinking about reproduction.

[19] Harris, 'Rights and Reproductive Choice', 35.

[20] Harris, 'Clones, Genes and Human Rights', claims that 'Article 16 if it is to be coherent at all must include the right of procreative autonomy', p. 93. The Article begins with the assertion 'Men and women of full age, without any limitation due to race, nationality or religion, have the right to marry and to found a family.' It does not define what a family must consist of, and indeed it is hard to see how a culturally neutral, universal conception of the family could be established. An assumption that couples without children are not families is gratuitous; many would find it offensive.

Appeals to individual autonomy, thought of as independence or self-determination or self-expression, can provide convincing arguments for the use of contraception. They will also, if the embryo or early foetus does not have the full rights of persons, provide reasons for legalising (early) abortion. In both of these cases the aim of the woman or couple involved is *not to reproduce*: there is no need to consider the rights, welfare or future of any child, since no child will exist. But where the aim is *to reproduce*, appeals to individual or personal autonomy are much less convincing. Reproduction aims to create a dependent being, and reproductive decisions are irresponsible unless those who make them can reasonably offer adequate and lasting care and support to the hoped-for child.[21]

Reproduction, whether by old or new methods, can never be justified simply by the fact that it expresses the individual autonomy of one or two (or more) would-be reproducers. An adequate future for children and their long dependence must aim to ensure that each child is born not just to an individual who seeks to express himself or herself, but to persons who can reasonably intend and expect to be present and active for the child across many years. Aspiring parent(s) will always find that what they can reasonably intend and expect is limited in quite prosaic ways. For example, although it is sometimes feasible in rich societies for a single individual to take on the full tasks of child rearing, any such project has to be judged against the reality that childhood is long and life uncertain, and that children need parents who are reliably present and active.[22] In rich and poor societies alike, those who are chronically ill or addicted, very young or very old, incapable or uncommitted cannot reasonably intend or expect to be active and present throughout a childhood. Nor in general can individuals without long-term and stable cohabitation and collaboration

[21] This point does not assume that we can individuate future persons, but only that we can specify them under some description or another: simple enough. Basic questions about obligations to those who will be born can be posed without metaphysical gymnastics.

[22] Failure to take these realities into account wholly undermines Shulamith Firestone's vision of procreative autonomy as requiring only short-term parental commitments. See Shulamith Firestone, *The Dialectics of Sex: The Case for Feminist Revolution*, William Morrow & Co, 1970.

with others reasonably expect to be present and active for the time and to the degree needed. Mere appeals to procreative autonomy, or more generally to self-expression or to individual autonomy, cannot make an otherwise questionable decision to reproduce ethically acceptable; nor can they establish any unconditional right to use particular reproductive technologies in pursuit of reproductive aims.

The literature on reproductive autonomy has startlingly little to say about parenthood or its real demands. Perhaps the individualistic focus of the favoured interpretations of autonomy blurs any but the most cursory consideration of the fact that the aim of reproduction is the creation of a long-dependent human person. Appeals to adult 'rights' to individual autonomy simply do not provide a sufficient ethical basis for reproductive decisions. The dependence of children provides good reasons to restrict the use of assisted reproductive technologies to those with adequate health and capacities, who have reasonable expectations and intentions of being active and present to bring up the child they aspire to bring into the world.

This conclusion may seem harsh in view of the fact that the readily fertile often reproduce without meeting these conditions. Failure to offer reproductive assistance to the infertile is sometimes criticised, as by Harris, as discriminating against them, so setting an unacceptable constraint on their (supposed) reproductive autonomy. Should we not, he suggests – and the point is widely made – view infertility as a disability, and seek to put the infertile in the same position as the fertile by providing them with full (even with free) access to reproductive technologies? In particular, would it not be discriminatory to place conditions on access to these technologies, given that the readily fertile are not prevented from having children just because they are ill or addicted, immature or incapable, or wholly unlikely to be committed parents?[23] I believe that this claim that regulating access to reproductive technologies invariably discriminates is strained.

[23] Harris, 'Rights and Reproductive Choice', 7.

It is no doubt true that many fertile persons have children without reasonable expectations or intentions of being active or present for those children, and even that some have children whom they do not much want, whom they later sometimes neglect or even abuse. This often leads to great harm, even to the removal of children from the demonstrably inadequate care of one or both parents, or to children suffering long periods of low-standard care. However, to forestall bad parenting except by means of education, encouragement and social support would require very serious infringements of the basic liberties of fertile persons who were prevented from having children. Short of coercive intrusions into the most basic liberties of the person – such as enforcing celibacy or forced abortion – there is nothing to be done until after a child is born. For example, if a woman conceives while addicted to heroin we rightly see this a misfortune for her child to be and for her: yet we do not use coercion to prevent such conceptions, or to force their termination, because this would be a grave violation of basic liberties of the person. We generally sanction coercive intervention only once a child is born, and then only if it is demonstrably abused or neglected. Infertile persons who – for whatever reasons – cannot reasonably expect or intend to be active and present for a child have, I think, an equal claim against coercive interference. But they have no extra claim to reproductive assistance, such as others' co-operation in or subsidy for any reproductive plan that may harm a foetus, an infant, or a child.

Boundary lines are hard to draw in these matters, but I think that it is reasonably clear that *by itself* an appeal to individual autonomy or to self-expression cannot establish a right to reproductive assistance for those who (for whatever reasons) lack the health, capacities and commitments for coping with the demands of parenthood. To restrict access to reproductive technologies to those who are fit, capable and committed to being at least adequate parents is no more discriminatory than restricting fostering or adoption to those with adequate health, capacities and commitments. In saying this, I do not suggest that reproductive assistance should be available only, let alone automatically, to married couples with excellent health and stable employment. As is well known, bad families may take just this form, and good families can take other forms. The test for those who seek assistance with reproduction, as for responsible aspiring parents who do not need reproductive assistance, is whether they are capable of and committed to being present and active for a child across a very large number of years.

3.6 THE LIMITS OF REPRODUCTIVE AUTONOMY

It is a truism that family life requires us to limit our individual autonomy, however conceived. An approach to the ethics of reproduction that centres largely on a conception of individual autonomy is therefore *prima facie* implausible. Recent debates on reproductive or procreative autonomy have not shown that individual or personal autonomy can be the sole, or even the central, ethical consideration in reproductive decisions.

Reproduction is intrinsically not an individual project. Nearly always it begins with a relationship – often imperfect, sometimes brief or even exploitative – between a man and a woman. Whenever it is successful it demands long-term relationships (whether biological or not) to protect and care for the child who has been brought into existence. Individual autonomy may appropriately have been centre-stage in early and mid-twentieth-century debates about limiting fertility and reproduction (although the appeal there was not strictly to reproductive autonomy, but to autonomy in all spheres of life). Here a right to choose can protect responsible reproductive decisions – for example, to have no children, fewer children or better-spaced children. However, even in this context, the thought that individual autonomy is important in reproductive decisions is not equivalent to the thought that there is a species of autonomy, namely reproductive or procreative autonomy, which is of special importance.

The visions of reproductive autonomy that have attracted so much attention in recent discussions are neither convincing nor attractive because they misconstrue having children as a form of self-expression that demands individual autonomy. Limiting child-bearing may promote individual autonomy, but having children generally curtails it.[24] Where appeals to individual autonomy play a central or decisive role in decisions to have children, the interests of the weak and vulnerable may be subordinated to the self-expression and preferences of the relatively powerful. It follows, I think, that appeals to individual autonomy cannot support unrestricted reproductive rights.

The advocates of procreative autonomy generally respond to this line of thought by deploying some version of Mill's classical views about the limitations of individual freedom, namely that the 'sole end for which mankind are warranted, individually or collectively, in interfering with the liberty of action of any of their number, is self-protection'.[25] They acknowledge that autonomy, including reproductive autonomy, may be constrained to prevent harm to others, including preventing harms to future children. Yet prevention, or avoidance, of harm is hardly enough in this case. The first task of parents is indeed not to harm: parents (and others) do wrong if they abuse or neglect a child in their care. But refraining from abuse and neglect is not even a minimally adequate standard for parents, whose central task is to care for and nurture the child.[26] Avoidance of harm is not a sufficiently robust constraint on individual autonomy in procreative decisions. The question to be asked is not just whether reproducing in certain situations, or using a particular technology, can sometimes be done without harming. It is whether there are reasonable grounds to think that any child brought into existence can expect to have at least an adequate

[24] The exception might be where progenitors see a large family as a guarantee of future prosperity and power. They may believe, for example, that many children will make a large labour contribution, or will continue their dynasty and its power.

[25] John Stuart Mill, *On Liberty* (1863), in *Utilitarianism, On Liberty and other Essays*, ed. Mary Warnock, Fontana, 1962.

[26] Thomas Murray, *The Worth of a Child*, University of California Press, 1996; David Archard, *Children: Rights and Childhood*, Routledge, 1993; Deech, 'Cloning and Public Policy', 95–100; Onora O'Neill, 'Children's Rights and Children's Lives', *Ethics*, 98, 1988, 445–63.

future, cared for by a 'good enough' family (biological or not) who will be present and active for the child across the long run.

Some reproductive practices that have been commended as acceptable expressions of individual autonomy 'provided they do no harm' clearly risk inadequate nurture and care of children who may be brought into existence. For example, enabling elderly women to have children whom they are unlikely to have the health and years to nurture well knowingly risks serious difficulties for any child.[27] The fact that other children with younger mothers may also lose their mothers too early in life is not a sufficient argument for choosing to have children very late in life: such loss is widely seen as grave misfortune, and hardly something for which even minimally responsible aspiring parents would plan.

Or again, the example of reproduction by cloning, which Harris sees as acceptable if it can be done safely, seems to me to be something for which no responsible parents would plan, even if safety issues were resolved (and they are far from resolved). It is not analogous to fertility treatments by which biological parents with remediable fertility difficulties become parents, where *from the child's point of view* difficulties resolved with his or her birth are immaterial.

Would-be parents by cloning, who use reproductive tissue and genetic material from themselves or their relatives, aim to bring into existence a child with *confused* and *ambiguous* family relationships. Family relationships are confused when *several individuals hold the role of one*; they are ambiguous when *one individual holds the roles of several*. What counts as 'confused' or 'ambiguous' will of course differ in different kinship systems. In Western societies it is standard to have only one mother, hence potentially confusing for a child when more than one person lays claim to the role (e.g. stepmothers, foster mothers). In some tribal societies it may be usual for a child to have a number of (social) mothers, and this need not be a source of confusion. Confused relationships are not invariably bad, but

[27] Contrast Harris, 'Rights and Reproductive Choice', with Inez de Beaufort, 'Letter from a Postmenopausal Mother', in John Harris and Søren Holm, eds., *The Future of Human Reproduction: Ethics, Choice and Regulation*, Clarendon Press, 1998, 238–47.

they are widely recognised as difficult for children, and for others. Where children acquire confused relationships, whether by fostering, adoption or parental remarriage, we usually see the situation as regrettable even if unavoidable, and seek to provide them with extra legal and social protection. Confused relationships created by cloning (or other technologies) are no less likely to burden children: would responsible parents seek confused relationships for their children from the start?

Ambiguous family relationships are less common, although they can arise by marriage between close relatives and by incest as well as by cloning. In such cases grandfather may be father, aunt may be sister. Cloning from a would-be parent is likely to produce more ambiguities even than incest within the nuclear family. Again responsible parents would not aspire for a child to have ambiguous relationships. Still less would any responsible parent plan for a child to have family relationships that are *confused* and *ambiguous*.

The only pattern of cloning that would produce genetic connection without ambiguous and confused relationships would be cloning from a child, dead or living, of the same couple: either possibility is once more one about which responsible aspiring parents might have reasonable misgivings. Responsible aspiring parents are therefore unlikely to favour cloning from themselves or from relatives: if it is unsafe it will fail or risk creating genetically damaged children; if it becomes safe it will create children whose sense of their descent and genealogy must be confused and ambiguous.

The problems of confused and ambiguous relationships could be avoided if would-be parents by cloning used oocytes and genetic material from strangers, so bringing into existence children without genetic link to themselves. But in this case they can achieve the same result without using complex or risky reproductive technologies, by adopting. (I leave aside the vulgar fantasy of cloning from a rich or famous third party.) Adoption is not the easiest route to parenthood, but it is well explored, socially accepted and generally regulated to forestall or manage anticipated difficulties and to support adopting parents. The difficulty of adoption is not moreover, as often suggested, a shortage of children for adoption, but rather the legal constraints on cross-border adoption: the world is full of children of all ages who desperately need a loving family.

The legal and regulatory standards that we impose on foster parents and adopting parents do not seem to me to represent an unacceptable restriction on procreative decisions or procreative rights, but rather a reasonable attempt to ensure that the needs of children for at least adequate (and preferably better than adequate) care and nurturing are not put at risk. Once we allow appeals to 'reproductive autonomy' no more than their due weight, there will also be reasons for regulating technologies for assisted reproduction.[28]

Between the reproductive technologies that are clearly immaterial from the child's point of view, and those whose use may create difficulties for a child, are others in which some familial relationships may be confused, although generally not ambiguous.[29] Reproductive technologies that use sperm or eggs or gestation supplied by third parties raise questions because they create families in which biological and social roles are not aligned. We have so far less complete knowledge than might be imagined of the effect of separating biological from social roles by using these reproductive technologies. Some empirical findings suggest that children produced by gamete donation and surrogacy can flourish, just as adopted children and stepchildren can flourish, but there is still reason for caution.[30] One reason for caution is that so far many children born of donated gametes have been left ignorant of the fact. Family secrets, as is well known, can have high costs; and so can the disclosure of family secrets. When and if these children

[28] One reason why the Human Fertilisation and Embryology Authority, which has regulated assisted reproduction in the UK since its creation in 1990, has earned respect from persons of many different views is that it has taken a careful approach while not confining treatment to any one type of family. For example it has not permitted use of PID for sex-selection (except to avoid genetic disease) and has indicated that it will take a firm stand against reproductive cloning.

[29] Ambiguity can also arise from sperm or egg donation within the family.

[30] See Rachel Cook, 'Donating Parenthood: Perspectives from Surrogacy and Gamete Donation', in Andrew Bainham, Shelley Day Sclater and Martin Richards, eds., *What is a Parent? A Socio-Legal Analysis*, Hart Publishing, 1999, 121–41.

discover the truth about their origins they may find it difficult and distressing. It may or may not be more problematic than that of adopted or stepchildren, where all concerned are usually well aware of the reassignment of parental roles, but where children face the costs of family breakdown.

Some evidence of the costs of late discovery about origins, unaccompanied by family breakdown, may be gleaned from the impact of unthought-through disclosure of DNA test results obtained to resolve suspicions about paternity. Yet where egg or sperm are donated, transparency from the start may also be risky. It would require either that children be told that they have a parent who disowns all contact and concern – or that donors or vendors of gametes take on a quasi-parental role. Perhaps we can imagine a society in which one of these is achieved without damage to children: in California children are now coming of age who will be entitled (if they have been told that this is how they were conceived) to ask for contact with the vendors from whom their mothers bought sperm.

3.7 REPROGENETICS AND PROCREATIVE AUTONOMY

Despite their initial attractiveness, many claims to reproductive autonomy are unconvincing in the context of real human relationships where children are vulnerable to adult pursuit of individual autonomy. However, the most impressive forms of reproductive autonomy may not be those that prevent or assist fertility, but those that use reprogenetic technologies to have genetically enhanced children: the 'designer babies' of countless media 'stories'.

Genetic technologies might contribute to individual autonomy in at least three substantial ways. In the first place, genetic therapies may enhance individual autonomy by alleviating or curing painful or progressive genetic diseases. Although large hopes are pinned on the development of therapies that will cure genetic diseases, there are so far few proven gene therapies. As such therapies are developed, their introduction and use will in general raise questions no different from those raised by the introduction of other new therapies, and their contribution to individual autonomy will not (I think) be different in kind from the contribution made by treatments for non-genetic conditions. (Germ line gene therapy might raise some distinctive issues; it is at present illegal in the UK and some other states.)

A second, much discussed, way in which genetics may contribute to human autonomy could be by genetic enhancement, i.e. by genetic modifications that produce human beings with different and supposedly enhanced capacities, including greater capacities for individual autonomy. Although levels of enthusiasm vary, there is already a substantial literature on the ethical issues of seeking to have genetically enhanced children. Attitudes range from the view that genetic enhancement is an aspect of parental reproductive autonomy, on a par with seeking to provide one's child with a good education,[31] to the view that it is acceptable provided it does not lead to unfairness or injustice,[32] to the view that it is a barbaric reversion to the old eugenics. If genetic enhancement permitted choice of children with a specified emotional or cognitive profile, a debate about genetic enhancement would be urgent. It would be urgent in part because technologies for genetic enhancement might or might not be used in ways that support individual autonomy: we have only to remember Aldous Huxley's *Brave New World* to see that enhancing individual autonomy is not the only possibility on the cards.

However, the largest contribution that genetics can currently make to human reproductive autonomy is of a third sort. Genetic information can be used to inform reproductive choices, and so to reduce births of children with severely disabling genetic diseases. Prevention of this sort does not add directly to anyone's individual autonomy: those who are never born do not thereby gain in individual autonomy. However, the use of genetic information to inform reproductive decisions may limit damage to the individual

[31] Harris, *Clones, Genes and Immortality*; Philip Kitcher, *Lives to Come: The Genetic Revolution and Human Possibilities*, Simon & Schuster, 1996; Lee M. Silver, *Remaking Eden: Cloning, Genetic Engineering and the Future of Humankind?*, Weidenfeld and Nicolson, 1998.

[32] Buchanan, Brock, Daniels and Wikler, *From Chance to Choice*.

autonomy of those who would otherwise have to care for severely disabled children, and may reduce the number of children born with little capacity even for minimal individual autonomy. Once a couple know that they risk having a child with a serious genetic disease they may choose to avoid doing so by various methods: they may choose not to have children; to use pre-natal testing with the intention of terminating affected pregnancies; to use donated sperm or eggs as relevant; to use IVF with PID. Any of these methods will prevent the birth of child with an identified risk of serious disease. So, too, can the use of social policies – prenuptial counselling, arranged marriages – to forestall the marriage of persons known to have recessive genes for the same serious disorder.[33]

Those who use currently available genetic technologies for reproductive purposes are not likely to see them as ways of enhancing their own or their children's individual autonomy. They are more likely to see them as useful for ensuring that their children do not suffer certain genetic illnesses, so at least have ordinary capacities for individual autonomy, however conceived. Like other parents, they will also realise that parenthood will reduce and restrict their own individual autonomy. Informed consent procedures are just as important in reproductive medicine as in other areas of medicine, but more robust and demanding conceptions of individual autonomy seem of remote relevance.

[33] Nuffield Council on Bioethics, *Genetic Screening: Ethical Issues*, 1993, see institutional bibliography; Andrew Bainham, Shelley Day Sclater and Martin Richards, eds., *What is a Parent? A Socio-Legal Analysis*, Hart Publishing, 1999, 121–41; Onora O'Neill, 'The 'Good Enough' Parent in the Age of the New Reproductive Technologies', in Hille Haker and Deryck Beyleveld, eds., *The Ethics of Genetics in Human Procreation*, Athenaeum Press, 2000, 33–48.

CHAPTER FOUR

Principled autonomy

4.1 THE FAILINGS OF INDIVIDUAL AUTONOMY

The claims of individual autonomy, in particular of patient autonomy and reproductive autonomy, have been endlessly rehearsed in bioethics in recent decades. By themselves, I have argued, conceptions of individual autonomy cannot provide a sufficient and convincing starting point for bioethics, or even for medical ethics. They may encourage ethically questionable forms of individualism and self-expression and may heighten rather than reduce public mistrust in medicine, science and biotechnology. At most individual autonomy, understood merely as an inflated term for informed consent requirements, can play a minor part within a wider account of ethical standards.

Even when individual autonomy is coupled with other ethical standards, problems persist. Most often it is combined with a Millian principle of avoiding harm. This is unsatisfactory. If we assume a full Utilitarian account of maximising happiness, we subordinate and marginalise individual autonomy itself; if we do not, the line between harmful and non-harmful action and policies will often be blurred. The supposed triumph of individual autonomy over other principles in bioethics is, I conclude, an unsustainable illusion.

The difficult and serious tasks remain. We need to identify more convincing patterns of ethical reasoning, and more convincing ways of choosing policies and action for medical practice and for dealing with advances in the life sciences and in biotechnology. In this chapter I shall argue that a different and older view of autonomy supports a more convincing approach to ethics, and also to bioethics. That other conception of autonomy was set out in Kant's writings. This may seem an unlikely source for a more convincing approach to bioethics. Many distinguished writers blame Kant for launching the unconvincing conception of individual autonomy, and stress its failings and blemishes.[1] They accuse Kant of identifying autonomy with self-control and independence, with extremes of individualism and with blindness to the ethical importance of the emotions and of institutions. The Kantian texts on which these readings of Kantian autonomy rely are well known; here I shall draw on a wider range of texts, but without setting out the full range of interpretive issues that have led me to the position I shall discuss. In this chapter I shall argue that Kant's distinctive conception of autonomy is quite different from the ethically inadequate conceptions of individual autonomy so commonly ascribed to him; that the links between Kantian autonomy and practical reason are strong and powerful; and that its ethical implications underpin the importance of trust. This may seem a perverse agenda, and before I embark on it I shall canvas some other possibilities, point to some problems they encounter and argue that it is worth taking the original Kantian approach to autonomy seriously.

4.2 HUMAN RIGHTS AS A BASIC FRAMEWORK?

Many see human rights as providing a promising framework for bioethics. Human rights supposedly provide good reasons both for serious respect for individual autonomy *and* for definite prohibitions on those uses of individual autonomy that violate others' rights. The approach will be promising if some way of identifying and justifying basic human rights claims can be found.

There are those who go no further in seeking justifications for rights than to appeal to the great charters and declarations, and in

[1] Accusers are numerous and distinguished. They include Iris Murdoch, *The Sovereignty of the Good*, Routledge & Kegan Paul, 1970; Bernard Williams, *Ethics and the Limits of Philosophy*, Fontana, 1985 and Simon Blackburn, *Ruling Passions: A Theory of Practical Reasoning*, Clarendon Press, 1998.

particular to the United Nations Universal Declaration of Human Rights of 1948.[2] These appeals are not philosophically very respectable because they amount to arguments from authority, but they are widely made and accepted. They are sometimes buttressed with claims that the status of the UN documents *even if initially they were mere declarations*, is now legitimated by their subsequent ratification by member states and by treaties agreed between states.[3] Cumulatively, it is suggested, the rights declared in 1948, and reiterated and expanded in subsequent UN and other documents have now been justified by the ratification that they have subsequently received from many governments. This may provide a good, or at least a middling, sort of argument to show that the human rights proclaimed in UN documents have been *politically legitimated*, but fails wholly as an *ethical justification* of those rights. Nor do reiterations, reformulations and extensions of those rights in subsequent documents of the same type – such as the European Convention on Human Rights and Biomedicine[4] – add one iota of ethical justification. One can see this in a negative way by noting that it is a purely procedural argument: treaties and other instruments with ethically rebarbative content can also be ratified by states, and can thereby gain political legitimation: but this would not make them ethically acceptable let alone ethically required. A variation of the argument looks more promising but is circular: if we presuppose states and democratic governance that are themselves ethically acceptable, ratification could (perhaps) confer ethical justification on the content of the documents ratified. But it is well known that any justification of state structures and democratic governance will be difficult, probably

[2] *Universal Declaration of Human Rights*, reprinted in Ian Brownlie, ed., *Basic Documents on Human Rights*, Clarendon, 1981, 21–7.

[3] Philip Alston, 'International Law and the Human Right to Food', in Philip Alston and K. Tomaševski, eds., *The Right to Food*, Nijhoff, 1984, 9–68.

[4] *Convention for the Protection of Human Rights and of the Dignity of the Human Being with regard to the Application of Biology and Medicine*, Council of Europe, DIR/JUR (96) 14, 1998; see institutional bibliography. If ratifications supplied justification many bogus and ill-defined 'rights', including a *ius primae noctis* and the supposed rights of slave-owners, could be justified.

impossible, without relying on a conception of justice, typically one incorporating an account of human rights. Advocates of human rights deny that states or governments are unconditionally just or justified. So have most people who have thought about justice since at least the late seventeenth century. But if the justice and justification of states and democratic governance *presupposes* the justification of human rights, human rights can hardly be justified by pointing out that they have been endorsed by states or by (elements of) democratic process. So much for one popular fantasy about hauling human rights into existence by their own bootstraps.

Bioethics is not a free-floating discipline: there is no way of justifying principles and standards by fiat or by proclamation, and no way of anchoring an account of human rights by mere appeal to declarations and charters, however august. Processes of ratification (by democratic states) may provide (some) democratic legitimation; they are not even qualified to provide ethical justification. This quick and lazy 'justification' of human rights fails, and a more strenuous approach is needed.

We need to think more carefully about justifying some basic principles, and to find reasonably secure arguments. It is highly plausible to think that any broader account of ethical justification, although it cannot begin with human rights, must sustain reasonably secure claims about some human rights. There are, I think, broadly two routes by which this might be approached.

4.3 GROUNDING HUMAN RIGHTS IN THE GOOD

The first and the more popular way of anchoring human rights claims ethically is to seek a convincing account of the good, of the good for man or of human interests, from which to derive an account of human rights. This way of looking at the matter is popular but difficult. Although we may be able find limited points of agreement about what is good – for example, agreement about a few basic human needs – agreement on a wider account of the good that is robust enough to sustain a systematic account of human rights has proved elusive.[5]

These disagreements would be resolved if we had a convincing account of the full range of possible human goods, such as an Aristotelian view of the good for man, or a neo-Aristotelian account of all the valuable functionings in human life.[6] I set aside as implausibly ambitious the fantasy that we shall find either a monistic account of the good, or a universal metric for all goods of the sort that would sustain any form or analogue of Utilitarian calculations. The difficulty is to supply *any* convincing arguments. How much of Aristotelian or other metaphysics would we need to provide a convincing account of the many components of a complete pluralistic account of the human good? How successful are those approaches that try to provide us with quasi-Aristotelian accounts of human good(s) without relying on traditional metaphysical arguments?

And supposing that we could achieve agreement on the many components of the human good, how could we move from it to an account of human rights? Claims about rights are systematic in a way that claims about goods are not: a pluralistic view of human goods is not a comfortable basis for a normatively coherent account of human rights. Just as a shopping list will not *in itself* contain information that requires some purchases to be given priority over others, so a pluralistic account of human goods does not *by itself* require some goods to be respected at the expense of others. It therefore provides no secure anchorage for rights claims. Rather than embark on the high seas of demanding arguments about the complete human good, and then about the priority of its components, I believe we are likely to reach a convincing account of human rights more directly by way of an account of human obligations.

[5] Onora O'Neill, *Towards Justice and Virtue: A Constructive Account of Practical Reasoning*, Cambridge University Press, 1996b.

[6] For a recent and very ambitious neo-Aristotelian account of the multiple elements of the human good see Martha Nussbaum, *Women and Human Development: A Capabilities Approach*, Cambridge University Press, 2000.

4.4 GROUNDING HUMAN RIGHTS IN HUMAN OBLIGATIONS

I see several advantages in seeking to anchor an account of human rights in an account of *human obligations* (or *human duties*), rather than in an account of the *human good.*

The first advantage is that obligations are structurally connected to rights; the second is that their connection to action can be well articulated; the third and consequential advantage is that obligations are more readily distinguished and individuated than are rights; the fourth is that the approach is less individualistic than rights-based approaches. Finally (and crucially) I believe that we can find better routes to the justification of obligations, and hence of rights, than we can find to the justification of rights, and hence of obligations.

The first advantage can be gained by making either obligations or rights fundamental. Unlike goods, rights and obligations are requirements, viewed respectively from the perspective of those who are to *receive* and of those who are to *act.* This structure of requirement links rights to their counterpart obligations[7]: they are alternative ways of looking at the same requirements. Any human right must have as its counterpart some obligation: a right that nobody is required to respect is simply not a right. Of course, we often find that the institutional basis for certain obligations and for the rights that correspond to them is lacking, or defective. When we do, we may wish to make claims about rights (and possibly about obligations) that, as it were, run ahead of any secure possibility of satisfying them. A proleptic rhetoric of rights (and possibly of obligations) can be politically valuable: in claiming rights that are not presently secured, in asserting obligations that ought to be secured, we may at least gesture towards securing them.

Yet a proleptic rhetoric has its dangers because it is too easy, all too easy, to proclaim illusory rights (there is rather less temptation

[7] This does not rule out the possibility of obligations without counterpart rights. See O'Neill, *Towards Justice and Virtue.*

to proclaim illusory obligations, making a proleptic rhetoric of obligations less problematic).[8] One unfortunate effect of the ease of proclamation can be found in hankerings for rights that are strictly impossible, or clearly incompatible with other more fundamental rights and obligations. I have sometimes heard a 'right' proclaimed that would be fundamental to bioethics – if it were not incoherent: I was taken to task at a WHO meeting for casting doubt on the supposed 'right to health'.[9] I had suggested that since it will never be possible to guarantee health for all, there can be no obligation so to do, and had concluded that there can therefore be no right to health. At the most, I suggested, we might argue for a right to health care of a certain level, with coherent corresponding obligations to provide that health care. My critic insisted that health is so important to human beings – which it is – that we should never in any circumstances cast doubt on the supposed right to health! Evidently he – and many others – did not view rights and obligations as requirements, but rather as ideals or aspirations. In my view this well-intentioned reinterpretation fatally weakens the entire point and purpose of talking either about rights or about obligations, because it does not construe them as requiring action: a disaster that will corrode and weaken all subsequent claims about rights and obligations.

Secondly, the performance of obligations is the business end of ethical requirements, because it is more closely and clearly connected to action. If nobody takes action to meet their obligations, nobody's rights will be respected. Simone Weil took the strong view that obligations – duties – are prior to rights when she began *The Need for Roots* with the words:

[8] Less tempting, but not unknown. Some writers claim that human obligations are either in principle or in practice unending. The most obvious examples are found in Utilitarian thought: see Peter Singer, 'Famine, Affluence and Morality', originally in *Philosophy and Public Affairs*, 1, 1972, 229–43, since much reprinted, and James Fishkin, *The Limits of Obligation*, Yale University Press, 1982. Are claims that obligations are in principle unending ethically serious – or the reverse?

[9] Since then the slogan *right to health* has been widely used, mostly (confusingly) to mean *rights to health care*, or *rights to health care of a specified standard.*

> The notion of obligations comes before that of rights, which is subordinate and relative to the former. A right is not effectual by itself, but only in relation to the obligation to which it corresponds, the effective exercise of a right springing not from the individual who possesses it, but from other men who consider themselves as being under a certain obligation towards him.[10]

This claim for the priority of obligations may well be correct. Theories of obligations antedate theories of rights; until recently accounts of rights did not fail to discuss obligations, whereas some older accounts of obligations had little to say about rights. However, for present purposes I shall claim not that obligations are prior to rights, but more circumspectly that rights and claimable obligations cannot be separated: they are the figure and the ground of one and the same moral pattern, views of a single set of requirements from two perspectives. An analogy may help. We might choose to describe a chessboard either as a white ground with black squares on it, or as a black ground with white squares on it. Neither the white nor the black squares are more fundamental to a chequerboard pattern. Rather they form, limit, indeed constitute one another. It would be absurd to ask whether the black or the white squares of a chequerboard pattern are more fundamental. Just as figure and ground are mutually determining, so rights and claimable obligations are mutually determining. It follows that where there are no obligations there are no rights.[11] This is hardly a novel claim. But in the last fifty years, this elementary point has unfortunately been submerged in a tumultuous enthusiasm for a disconnected rhetoric of rights that overlooks the damage done to rights by failing to take their intrinsic connection to obligations seriously.

[10] Simone Weil, *The Need for Roots: A Prelude to a Declaration of Duties toward Mankind* (1949), trans. A. F. Wills, Routledge & Kegan Paul, 1952, 3.

[11] This point is not the same as that disputed between those who think that responsibilities go with rights, and those who do not. The thought that if *A* has rights then *A* has some responsibilities is plausible; but it is not a conceptual truth. It is quite coherent for some people to have only rights and no responsibilities (e.g. small children, the mentally disturbed). It is not even coherent to think that anyone can have rights where nobody has obligations to respect them.

The third advantage arises from the fact that we speak about obligations in the language of action, so can more readily individuate and distinguish obligations than we can rights. An obligation is an obligation *to do* or *to desist*, *to act* or *to refrain* in this or that way, in these or those situations, towards these or those others. But we all too often talk about rights using a substantival vocabulary: we speak of *rights to life*, or *rights to health care* (or *to health*!), or *rights to food* or *rights to privacy*, or *rights to choose*, as though rights were best thought of in abstraction from action as entitlements to entities or goods of one or another sort. The substantival rhetoric has its uses – often political uses – but it disguises real and practical questions more than it illuminates them. For these rights can be respected and secured only if some or many agents are obliged to act in specific ways towards certain others: for example if various agents are obliged to *provide (organise, pay for, deliver)* specific sorts of health care for specified others, or to *provide* (*grow, pay for, deliver*) specific sorts of food or funds to specified persons who lack food, or to *make available, fund, distribute or prescribe* contraceptive services for specified persons.

This problem is confounded by the fact that when we use a substantival rhetoric of rights it is very natural to conflate distinct rights in ways that the more active vocabulary of obligation is less likely to suggest. For example, the slogan *right to life*, quite apart from its role in abortion debates, has been interpreted as covering everything from an obligation not to kill others without reason (reasons variously thought to include just war, defence of the innocent, self-defence and (in some quarters) capital punishment) to an obligation to provide the means of life to those without means. The slogan *right to work* has been variously seen as a right not to be excluded from seeking employment, and as a right to a job. I quail even at the thought of counting how many interpretations have been proposed for the notion of a *right to equal opportunity*. To achieve clarity about rights, it is nearly always helpful to shift to the vocabulary of action, while retaining the normative vocabulary of requirement. In doing so we deploy the vocabulary of obligation.

Although the rhetoric of rights dominates public life today, I believe that we fail to take rights seriously unless we link rights claims to rigorous thinking about obligations, about action and about the capabilities that agents and institutions need in order to discharge their obligations, and thereby respect one another's rights. If we try to think about rights in isolation, we end up gesturing towards supposed rights by offering hazy indications of what they would secure, while neglecting to specify the action needed to respect them. This lazy approach risks bringing the entire enterprise to which the human rights movement subscribes into contempt.

Fourth, in speaking about obligations we shift from a way of thought that is often viewed (both by protagonists and by opponents) as individualistic, to one that takes *relationships* between obligation bearers and right holders, including institutionally defined relationships, as central. When we talk about rights, particularly in substantival ways, it is easy to imagine that we are talking about individuals: the claimant alone is in the frame, making rights claims against some unspecified other, or even against the world at large. Individual autonomy fits cosily into views that give priority to rights, and some hope to justify rights solely or largely on the basis of their supposed contribution to individual autonomy. But when we talk about obligations we immediately have to focus on relationships between obligation bearers and right holders, between obligation bearers and beneficiaries. We do not know what our obligations are if we cannot specify to whom we owe them (or, in the case of obligations without specified claimants, what types of action are needed if they are to be well carried out). Agents may have some obligations to all others, such as obligations not to torture or not to enslave; other obligations to individual persons, such as obligations to keep promises to those to whom they were made; and other obligations to persons who are not individuated but merely specified by one or another description – for example, to other road users, to nearby others in present danger, to neighbours or colleagues. These distinctions are bracketed and very easily obliterated if we emphasise rights at the expense of obligations.

4.5 KANT AND PRINCIPLED AUTONOMY

The advantages of grounding rights in obligations can be secured only if good arguments for central human obligations can be found. This has been the shared ambition of considerable parts of the modern version of the Natural Law tradition, and of its contemporary descendants, which seek accounts of justice that begin with obligations. In this enterprise Kant's work remains the most significant; it is also notable for the central role he assigns to autonomy. His view of the importance of autonomy to morality is powerful and uncompromising. He writes, for example, that '*Autonomy* of the will is the sole principle of all moral laws and of duties in keeping with them'.[12] If Kant had accepted the conception of individual autonomy that has enjoyed such a vogue in recent decades, especially in bioethics – which is often ascribed to him – this claim would be highly implausible. But he did not.

The evidence that Kant's account of autonomy is not a conception of individual autonomy is easily assembled.[13] He never speaks of an *autonomous self* or *autonomous persons* or *autonomous individuals*, but rather of the *autonomy of reason*, of the *autonomy of ethics*, of the *autonomy of principles* and of the *autonomy of willing*. He does not see autonomy as something that some individuals have to a greater and others to a lesser degree, and he does not equate it with any distinctive form of personal independence or self-expression, let alone with acting on some rather than other sorts of preferences. Kantian autonomy is manifested in a life in which duties are met, in which there is respect for others and their rights, rather than in a life liberated from all bonds. For Kant autonomy is *not relational*,

[12] Immanuel Kant, *Critique of Practical Reason* (1785), trans. Mary J. Gregor, in *Kant, Practical Philosophy*, Cambridge University Press, 1996, 5:33, 19–16; all the emphasis in the Kant quotations is in the original.

[13] Thomas E. Hill, Jr., 'The Kantian Conception of Autonomy' in Hill, *Dignity and Practical Reason in Kant's Moral Theory*, Cornell University Press, 1992, 76–96, discusses ways in which contemporary conceptions of autonomy have 'been cut loose from their Kantian roots', p. 77. See also Onora O'Neill, 'Autonomy and the Fact of Reason in the *Kritik der praktischen Vernunft*, 30–41', in Otfried Höffe, ed., *Immanuel Kant, Kritik der praktischen Vernunft*, Akademie Verlag, forthcoming; 'Kant's Conception of Public Reason', in *Proceedings of the IX Kant Kongress*, de Gruyter, Berlin, forthcoming.

not graduated, not a form of self-expression; it is a matter of acting on certain sorts of principles, and specifically on principles of obligation. From the perspective of contemporary fans of individual autonomy, whose work has almost obliterated the older Kantian conception of autonomy, this must seem far-fetched. But is it?

'Kant', J. B. Schneewind has claimed, 'invented the conception of morality as autonomy', the claim that 'we ourselves legislate the moral law'.[14] Kant expressed this point in the so-called *Formula of Autonomy*, one of a number of versions of the Categorical Imperative, in the words 'choose only in such a way that the maxims of your choice are also included as universal laws in the same volition'.[15] Familiar as the Kantian rhetoric is, this claim sounds bizarre. What on earth can Kant have meant by speaking here and elsewhere of '*the will of every rational being as a will giving universal laws*'? Is it even coherent to speak of each of a plurality of individuals who share a common world as 'universal legislators'? Isn't the metaphor of *legislation* tied firmly to the picture of an *individual legislator* or an *individual legislature*, that is to a *single* authority with an integrated decision-making procedure that produces a *single* set of laws to regulate the lives of a *plurality* of subjects within some domain?

Kant, I believe, understood his 'invention' of autonomy as universal self-legislation in a way that resolves this puzzle. His focus was not on any special sort of act of choice, by which each *actually* chooses laws or principles for everyone else, but on a distinctive constraint or requirement, a test that shows which principles of action *could* be chosen by all, that is to say which principles are universalis*able*, or *fit to be universal laws*. As Kant sees it, individuals can choose to act on principles that meet or that flout the constraints set by the principle of autonomy, but have reasons to act only on those principles that meet those constraints.

[14] J. B. Schneewind, *The Invention of Autonomy: A History of Modern Moral Philosophy*, Cambridge University Press, 1998, 3, 6.

[15] Immanuel Kant, *Groundwork of the Metaphysic of Morals* (1785), trans. Mary J. Gregor, in Kant, *Practical Philosophy*, Cambridge University Press, 1996, 4, p. 440. See Allen W. Wood, *Kant's Ethical Thought*, Cambridge University Press, 1999, 163–4, for a listing of versions of the Formula of Autonomy.

This modal conception of action on autonomous principles is clearly wholly unlike any contemporary conception of individual autonomy. The agent who acts with principled autonomy has indeed to be free to act, but need not have or show an unusually high degree of psychological independence. In tempting and difficult circumstances a little individual autonomy may be helpful for acting with principled autonomy – but large doses of individual autonomy may lead agents to flout principled autonomy. Nor are those who act with principled autonomy committed to any distinctive form of individualism: principled autonomy is expressed in action whose principle *could be adopted by all others*. Any conception of autonomy that sees it as expressing individuality – let alone eccentricity – or as carving out some particularly independent or distinctive trajectory in this world is a form of individual rather than of principled autonomy.

However in other ways principled autonomy is more demanding than individual autonomy. For Kant the term *autonomy* or *self-legislation* (*Selbstgesetzgebung*) is not, I believe, to be understood as legislation by an individual agent or self, an idea that expresses an extreme (probably incoherent) version of individual autonomy. A better reading of Kant's idea of *self-legislation* would view the element *self* in *self-legislation* simply as a reflexive term. Self-legislation is legislation that does not refer to or derive from anything else; it is *non-derivative legislation*. Legislation or principles derived for something else – for example, from a dictator's edict, or an individual's desire – lack justification unless that from which they are derived is justified. As Kant sees it, fundamental ethical principles should presuppose merely what it takes to be a principle at all. He therefore bases ethical reasoning on the ideal of living by principles that at least *could* be principles or laws for all, that have the *form of law*. Principled autonomy has nothing to do with especially tremendous or magnificent individuals who legislate for all; it has to do with especially significant principles or 'laws' that *can* be adopted by any, hence by all, ordinary agents. Kant's concern is not a *self* that actually legislates for all, but principles that are fit to be laws for all. The stress he places on the term *self-legislation* is on the notion of *legislation*: the advocates of individual autonomy by contrast stress the notion *self* and have little to say about any conception of (moral) legislation.

4.6 PRINCIPLED AUTONOMY AND HUMAN OBLIGATIONS

It is easy to imagine that principled autonomy will provide only a trivial constraint. Surely if a principle can be adopted by one person, then it can be adopted by any, and so by all? (The point after all is not to find principles that everybody can act on at all times or places – a common misconception of Kant's view – but only to identify principles that can be adopted by all, that any agent can 'will as a universal law'.) However, Kant's arguments are designed to show that this is not the case, by establishing that a range of very fundamental principles *cannot* be 'willed as a universal law': those who adopt them find that they cannot coherently will (even hypothetically) that all others adopt the same principle.

The demand that we act only on principles that we can view as principles for all is the basis for a rich and powerful account of ethical requirements. Kant works out these requirements by considering which principles cannot be willed as principles for all, whose rejection will therefore be obligatory for all. This would be trivial if every principle agents can will or adopt could be willed for all. However, Kant insists that willing is not mere wishing, so that willing as a universal law is not merely a matter of *formulating* a universalised principle with the same content as one's own proposal for action. In willing a principle of action we commit ourselves to take any necessary and some sufficient means, taking account of the reasonably foreseeable results of the action. In consequence, a range of tempting principles *cannot* be 'willed as universal laws': their principled rejection identifies the central ethical obligations, including the central obligations of bioethics.

For example, an agent who adopts a principle of coercion must also will some effective means of coercion (violence, intimidation, whatever else might work). So an agent who (hypothetically) wills a principle of coercion as a universal law must also (hypothetically) will that everybody use some effective means of coercion. However, since there will be at least some coercive action in any world where all are committed to a principle of coercion, at least some persons would then be unable to adopt a principle of coercion because their capacities for action would be destroyed or undermined or bypassed by others' coercive action. The reasonably foreseeable result of anything approaching universal commitment to coercion would ensure that there could not be universally available effective means to coerce: universal coercion is therefore an incoherent project. Coercion is necessarily a minority pastime, and universal coercion cannot be willed without internal contradiction.

Coercers therefore do not in fact suppose that universal coercion would be coherent. Rather, as Kant point out, they make an exception for themselves:

> If we now attend to ourselves whenever we transgress a duty we find that we do not in fact will that our maxim should become a universal law – since this is impossible for us – but rather that its opposite should remain a law universally: we only take the liberty of making an exception to it for ourselves or even just for this once.[16]

It follows that *if* we are committed to adopting basic principles that we could will others to adopt, we will have reason to reject a principle of coercion. It does not follow that all coercion is unjustified, for it may be that the best we can do if we are serious about rejecting coercion in so far as possible will still require some institutions that use certain, limited, regulated forms of coercion – for example, a police force, a taxation system. But these specific, limited, regulated uses of coercion would be justified only to the extent that they were indispensable elements of an underlying project of rejecting coercion and respecting other equally fundamental obligations.

There are many other principles of action that cannot be universal laws. Killing and coercing, injury and violence, manipulation and deception, torture and intimidation, enslaving and forced labour are all principles that cannot be willed as universal laws:

[16] Kant, *Groundwork of the Metaphysic of Morals*, 4:424.

those who seek to act on these principles cannot coherently will that everybody else do the same.[17] Putting the matter generally, any principle of action whose universal adoption would destroy, damage or undermine capacities for action for some or for many cannot be willed as a universal law. The rejection of principles that cannot be principles for all is, on Kant's view, the basis of human duty.

The argument so far might suggest that Kantian autonomy rejects only principles of destroying, damaging or undermining others' capacities for action, and their expression in policies, institutions and action. If so, his arguments would establish only those obligations that (roughly) correspond to liberty rights. However, Kant thinks that a variant of the same line of argument also justifies requirements to support and assist others. Vulnerable agents – and all human agents are vulnerable – cannot will indifference to others as a universal law because they invariably have plans and projects that they cannot reasonably hope to achieve without others' support. In willing that indifference should become a universal principle, they would (incoherently) will to put at risk help that may be indispensable for others' activities or projects, including their own. If (*per impossibile*) we try to make indifference a universal principle we commit ourselves to ways of acting and living that put everyone's (including our) own survival and quality of life at risk. Since universal indifference would be enough to destroy, damage or undermine human agency for many, willing a principle of indifference as a universal law is incompatible with commitment to seek effective means for whatever projects we seek to pursue.

The rejection of indifference will demand a lot. Although it cannot require any individual to render *all* needed assistance to *all* others in *all* predicaments (an impossibility), it demands far more than the sporadic meeting of others' needs, for example, by episodic charitable donations or emergency aid. As is well known, the capacities for action of those who depend only on such minimal, erratic sources of support are constantly at risk, frequently damaged and often obliterated. Those who reject indifference will

[17] In chapter 5 I shall trace the derivation of an obligation to reject deception, and its importance for trustworthiness and relevance to trust.

in most circumstances find that their commitment is best expressed by supporting social and political institutions and practices that reliably reduce and limit vulnerability by providing a reliable degree of security and subsistence for all, for example by arrangements that help make food and health care affordable and the environment safe.

These core Kantian arguments show that principled autonomy can identify substantial basic ethical requirements. But they show neither why we should take principled autonomy seriously nor what more specific rights and duties and what sorts of institutions are required. The argument needs extending in both directions from the familiar central ground of Kantian ethics. I shall begin with a brief account of the authority of principled autonomy. Although it is philosophically more demanding than other parts of this book, it is not counter-intuitive. In later chapters I shall consider what principled autonomy might contribute to reorienting bioethics and to developing a plausible account of trust.

4.7 TAKING PRINCIPLED AUTONOMY SERIOUSLY

Why should we take principled autonomy, the idea of acting on principles that we can will as universal laws, at all seriously?[18] Allen Wood puts this question pointedly in his work, *Kant's Ethical Thought*. He suggests that the idea that Kantian autonomy provides the basis and justification of moral obligation will fail on either of two tempting interpretations. He writes:

> Autonomy of the will as the ground of moral obligation is arguably Kant's most original ethical discovery (or invention). But it is also easy to regard Kant's conception of autonomy as either incoherent or fraudulent. To make my own will the author of my obligations seems to leave both their content and their bindingness at my discretion, which contradicts the idea that I am *obligated* by them. If we reply to this objection by emphasising the *rationality* of these laws as what binds me, then we seem to be transferring the source of obligation from my will to the canons of

[18] For textual background to this section see the papers referred to in n. 11.

rationality. The notion of *self*-legislation becomes a deception or (at best) a euphemism.[19]

Advocates of individual autonomy choose the first horn of Wood's dilemma. Wood's comment make its defects plain: 'To make my own will the author of my obligations seems to leave both their content and their bindingness at my discretion.' Conceptions of individual autonomy, whether read voluntaristically as mere, sheer pursuit of preference or more cautiously as pursuit of the right sorts of preferences (those that are second-order, reflectively endorsed, etc.) substitute self-expression for moral obligation. It may be a little brutal to speak of this conception of the source of obligations as incoherent, as Wood does; but individual autonomy (however interpreted) is singularly ill-suited to provide the basis for any account of obligation. Yet the second horn of Wood's dilemma may seem equally unpromising: 'by emphasising the *rationality* of these laws as what binds me, we seem to be transferring the source of obligation from my will to the canons of rationality. The notion of *self*-legislation becomes a deception or at best a euphemism.' If Kant derives the significance of universal legislation 'from the canons of rationality', he retreats (Wood suggests) to some form of intellectualism or perfectionism: he grounds morality not in principled autonomy but in some antecedently given conception of the good or of reason.

Kant countered this thought with the striking claim that *autonomy (principled autonomy!) itself is the fundamental principle of reason.* He puts this arresting thought very openly at times:

> The power to judge autonomously – that is, freely (according to principles of thought in general) – is called reason.[20]

Kant avoids the second horn of Wood's dilemma by claiming that there is no possibility of retreat to antecedently given

[19] Wood, *Kant's Ethical Thought*, 156, emphasis in the original.

[20] Immanuel Kant, *The Conflict of the Faculties* (1798), trans. Mary J. Gregor and Robert Anchor in Kant, *Religion and Rational Theology*, eds. Allen W. Wood and George di Giovanni, Cambridge University Press, 1996, 7:27.

canons of reason. He does not seek to derive principled autonomy from antecedently given standards of reason, *because he holds that we know of no such standards.* (Those book titles *Critique of Pure Reason*, *Critique of Practical Reason* gave fair warning!). The project of deriving morality from antecedently known standards of the good or from antecedently given standards of reason is a non-starter for Kant: reason, as he famously said, 'has no dictatorial authority'.[21] We cannot appeal to authority to discover the foundations of morality. If we try to do so, we reach only a simulacrum of morality that subordinates it to something else, illustrating the invariable limits of arguments from authority.

Despite Kant's great reputation, despite his explicit commitment to a critical view of reason, his writings on the authority of reason have been surprisingly but thoroughly neglected. We may approach the topic in an intuitive way by asking why *any* specific ways of thinking and acting should be thought of as having a general authority that would lead us to speak of them as *reasoned*, while others do not. Kant's fundamental thought is rather simple: reasons are the sorts of things that we give and receive, exchange and refuse. If anything is to count as reasoned it must be *accessible to others who are to be the audience for that reasoning.* Others must be able to follow, so find intelligible, speech or writing that is to count as offering reasons for thinking or believing; others must be able to adopt proposals for action made to them if these are to count as offering reasons for action. We do not offer others reasons for believing if we communicate with them in ways that we know they will find unintelligible; we do not offer others reasons for acting if we present them with proposals for action that we know they will find unadoptable. To make either thinking or proposals for action followable by others we must impose structure on them: we must make our underlying principles of communication and proposals for action that are law-like or principled; if we do not, we do not offer reasons for thinking or for acting.

[21] Immanuel Kant, *Critique of Pure Reason* (1781), trans. Paul Guyer and Allen W. Wood, Cambridge University Press, 1998, A738/B766.

For Kant the fundamental problem of offering an account of reason is that we do not find canons of reason inscribed antecedently in our minds. As he sees it, reason is not, as Descartes supposed, 'whole and complete in each of us'[22] it is not grounded by theological or metaphysical arguments. Its fundamental requirements are therefore no more the necessary conditions that anyone who seeks to reason with others must adopt. As Kant sees it, principled autonomy is no more – but also no less – than a formulation of these basic requirements of all reasoning. His tough and dramatic thought is that the demands of reason in theory and practice, in thinking and in willing, run parallel. Both are constituted or constructed by the specific, yet minimal structure that *must* be imposed on thought and action by each member of any plurality of agents who *can* follow one another's thinking and one another's reasons for action. The Categorical Imperative, in all its versions, including the Formula of Autonomy, articulates this double modal structure of the supreme principle of reason for the domain of action: we *must* act on principles others *can* follow. So there is no gap between reason and principled autonomy, and specifically no gap between practical reason and principled autonomy in willing.

Although this striking claim may initially seem implausible, Kant thought that it could be presented in ways that made it both accessible and appealing. His conception of principled autonomy combines two points. First, he needs to show that reasoning could not be a matter of *deference* to any antecedently given standard(s) (of some alleged but unvindicated authority). Second, he needs to show that reasoning also could not be a matter of lacking all standards, of *anomie.* The conclusion that reasoning is a matter of following non-derivative standards, hence only of the necessary requirements for being accessible to others, is then easily reached.

[22] René Descartes, *Discourse on the Method of Rightly Conducting One's Reason and Seeking the Truth in the Sciences, Philosophical Writings of Descartes* (1637), vol. 1, trans. John Cottingham, Robert Stoothof and Dugald Murdoch, Cambridge University Press, Cambridge, 1985, 112.

Intuitively easy versions of these three steps can be found in two popular essays of the mid-1780s, *What is Enlightenment?* and *What Does it Mean to Orient Oneself in Thinking?*[23] Each essay includes a *reductio ad absurdum* of the view that standards of reason are antecedently given, and simply require our allegiance or obedience to a supposed authority. Reasoning, Kant points out (rather plausibly), cannot be a matter of deference, for example to civil, ecclesiastical or popular views (or we may add to public opinion, contemporary discursive practices or citizens' juries). All forms of cognitive deference are no more than arguments from putative authority. They justify only insofar as independent reasons are offered for accepting that putative authority.

The second essay – *What Does it Mean to Orient Oneself in Thinking?* – extends the account of reason by arguing that non-deference cannot be all that there is to reasoning. Any use of our cognitive capacities that is not structured or disciplined at all, that is literally 'lawless' in that it incorporates no standards or principles, must fail because it provides nothing for others to follow. Agents who dispense with all standards or constraints, whose thinking and acting has no structure, are literally uninterpretable by others, perhaps ultimately to themselves: they may not defer to arbitrary standards, but their communication will be paralysed by lack of structure. Such agents simply fail to offer one another reasons either for thinking in certain ways or for acting in certain ways. Their supposed processes of thought and their supposed reasons for action are *anomic* or *lawless* rather than reason-giving, hence law-like and followable by others. (Kant condemns what we now call postmodernism as 'lawless' thinking that is ultimately not liberating but stultifying, a self-indulgent fantasy about the dispensability of all standards and principles in thinking and acting. Such 'lawless' thinking leads not to freedom of thought, not to liberation, nor to

[23] Immanuel Kant, *What Does it Mean to Orient Oneself in Thinking?* (1786), trans. Allen W. Wood and George di Giovanni in Kant, *Religion and Rational Theology*, Cambridge University Press, 1996, 8:133–46; Immanuel Kant, *What is Enlightenment?* (1784), trans. Mary J. Gregor, in Kant, *Practical Philosophy*, Cambridge University Press, 1996, 8:35–42. Kant, *The Conflict of the Faculties* 7:5–116, is also relevant to these themes.

the free flight of genius (as its varied proponents fondly fantasise) but to ultimately to incoherence and isolation.[24])

Putting the two steps of Kant's argument together, we easily reach the conclusion that the basic requirements for reasoning about what we are to think and what we are to do cannot be derived from elsewhere, yet must impose structure. If we are to reason we must think and act in ways that manifest principled autonomy rather than deference or randomness. Whatever views we reach about more specific types of inference or argument, this basic demand cannot be avoided or rejected. Turning away from the demands of principled autonomy amounts either to trying to 'vindicate' claims by introducing arbitrary premises – to which all conclusions must then be relativised – or to dispensing with any conception of reason. Either way, the cost is the loss of any prospect of offering reasons to others.

So reasoning, on Kant's account, is simply a matter of striving for principled autonomy in the spheres of thinking and of action. Autonomy in thinking is no more – but also no less – than the attempt to conduct thinking (speaking, writing) on principles on which all others whom we address could also conduct their thinking (speaking, writing). Autonomy in action is no more – but also no less – than the attempt to act on principles on which all others could act.[25] Wood's dilemma is avoided because Kantian autonomy is neither derived from an antecedently given but unjustified account of reason (hence unreasoned), nor lacking in structure (hence wilful and arbitrary): principled autonomy itself supplies the basic structures of reasoning.

[24] Kant, *What Does it Mean to Orient Oneself in Thinking?*, 8:146–7.

[25] 'Freedom in thinking signifies the subjection of reason to no laws except those which it gives itself; and its opposite is a lawless use of reason . . . if reason does not wish to subject itself to the laws it gives itself, it has to fall under the yoke of laws given by another; for without any law, nothing, not even nonsense can play its game for long. Thus the unavoidable consequence of declared lawlessness in thinking (of a liberation from all the limitations of reason) is that freedom to think will ultimately be forfeited . . . [or] . . . trifled away.' Kant, *What Does it Mean to Orient Oneself in Thinking?*, 8:303–4.

So 'self-legislation' is not a mysterious phrase for describing merely arbitrary ways in which a free individual might or might not act. It is the basic characteristic of ways of thinking or willing that are conducted with sufficient discipline to be followable by or accessible to others. Such ways of thinking and acting must be *lawlike* rather than *lawless*, and will thereby be in principle intelligible to others, and open to their criticism, rebuttal or reasoned agreement.

4.8 PRINCIPLED AUTONOMY, OBLIGATIONS AND RIGHTS

This freehand sketch of Kant's reasons for thinking that principled autonomy is a non-derivative, fundamental requirement on thought and action provides reasons for taking seriously the moral principles that can be derived from it. If we take these basic principles of morality – for example, the rejection of injury, of coercion, of slavery, of indifference – we shall need seriously to work out how people can live up to them. I do not think that we could expect to find any *timeless* account of the more narrowly specified human rights and human obligations that would express and implement these principles, or of specific institutional structures needed to realise these rights and obligations at all times and places. Nor do we need to do so. The task is rather to identify ways of living up to these basic principles in actual circumstances, with their historically contingent but determinate configuration of human and technical capabilities, material resources and environmental constraints. Equally in bioethics the task will be to identify ways of living up to these principles in actual circumstances, with their historically contingent but determinate configuration of medical, scientific and biotechnological resources and environmental constraints. In the following chapters I shall outline some of the implications, starting with an ethically serious conception of principled autonomy.

CHAPTER FIVE

Principled autonomy and genetic technologies

5.1 BEYOND INDIVIDUAL AUTONOMY

Individual autonomy, as we saw in chapters 2 and 3, offers an unsatisfactory approach to many ethical issues that arise in medicine, science and biotechnology. The minimal interpretation of individual autonomy as informed consent provides plausible but very incomplete ethical guidance; more robust interpretations of individual autonomy offer more complete but very implausible ethical guidance. Nor, as disputes about 'reproductive autonomy' show, can ethical issues be well resolved merely by limiting the pursuit of individual autonomy by a requirement not to harm.

But will ethical arguments that invoke principled rather than individual autonomy prove either more plausible or more complete? There are at least some reasons to hope that they may. Principled autonomy requires that we act only on principles that can be principles for all; it provides a basis for an account of the underlying principles of universal obligations and rights that can structure relationships between agents.

A primary focus on interaction and relationships, on obligations and rights, does not prevent those committed to principled autonomy from assigning due – but no more than due – weight to individual autonomy. Without some capacities for and some use of individual autonomy (variously interpreted) agents will lack the resolution and the self-confidence to fulfil their obligations and to respect one another's rights. Acting with principled autonomy needs a modest capacity for individual autonomy; but that necessary minimum is only one minor aspect of principled autonomy.

5.2 PRINCIPLED AUTONOMY, DECEPTION AND TRUST

Principled autonomy is a powerful basis for ethics because it can establish a number of distinct and fundamental principles of obligation. In chapter 4 I argued that a commitment to principled autonomy requires us to reject both coercion and deception, set out some arguments for rejecting coercion in some detail and indicated that there were parallel reasons for viewing the rejection of deception as a fundamental ethical requirement. Rejecting and avoiding coercion and deception are of ubiquitous and fundamental importance in ethics, and specifically in bioethics. One advantage of taking them seriously is that, taken together, they provide the basis for informed consent requirements: action that either coerces or deceives others stands in the way of free and informed consent; conversely where free and informed consent is given, agents will have a measure of protection against coercion and deception.

However, the importance of commitments to shun avoidable coercion and deception goes far beyond the role that they jointly play in justifying informed consent requirements. Together they provide the basis for many aspects of requirements to respect persons and for conceptions of confidentiality. Above all they provide reasons for seeking to establish, maintain and respect trustworthy institutions and relationships. Whereas individual autonomy is constantly in tension with relations of trust, principled autonomy provides a basis for relations of trust.

Relations of trust require us to reject deception just as they require us to reject coercion. This is readily appreciated when we note that deception is often a tempting and useful strategy, yet wreaks havoc in the lives of its victims. Deception offers covert ways of obtaining advantage or avoiding detriment; it is not always difficult; it is not always detected; even when detected, it is often another day and advantage has already been gained. So it is of great significance to establish a fundamental human obligation to reject deception. This obligation provides the ethical basis for trustworthy action; and trustworthy action can provide important evidence for anyone who seeks to place trust.

The basic argument for an obligation to reject deception parallels the argument for an obligation to reject coercion. Nobody who is committed to principled autonomy can make deception of others basic to his or her life and action because deception *cannot* serve as a principle for all. It cannot do so because one standard effect of widespread, let alone universal, deception would be severe and widespread damage to trust. This damage to trust would undercut or damage an indispensable prerequisite of any deception: if a principle of deception is widely, let alone universally, adopted some – or many – people will find that others will not accept their words or deeds as trustworthy, so that they are unable, or less able, to deceive. Deception cannot therefore be a principle of action for all: the rejection of deception is the underlying principle for a wide range of human obligations.

A commitment to reject deception will have many implications. It will be expressed in refraining from lying, from false promising, from promise breaking, from misrepresentation, from manipulation, from theft, from fraud, from corruption, from passing off, from impersonation, from perjury, from forgery, from plagiarism and from many other ways of misleading. More positively, it will be expressed through truthful communication, through care not to mislead, through avoidance of exaggeration, through simplicity and explicitness, through honesty in dealing with others, in a word, through trustworthiness.

Those who reject deception will not, however, have exceptionless obligations not to deceive or to be completely open in each and every circumstance. Just as some coercion (a police force, a tax system) must be accepted even by those whose fundamental principle is to reject coercion, so some forms of deception (habits of civility, toleration of 'white' lies, silence and discretion) must be accepted even by those whose fundamental principle is to reject deception. Although a commitment to principled autonomy does not entail an exceptionless obligation to openness in all possible circumstances, it requires strong commitment to honest and trustworthy action and communication. It is therefore of great ethical importance for all action, policies and institutions, and nowhere more so than in medicine, science and biotechnology. These points can be well illustrated by considering some ethical issues that are prominent in discussions of various genetic technologies.

5.3 GENETIC TECHNOLOGIES

In many parts of the world genetic technologies in current or incipient use are widely mistrusted.[1] Mistrust is directed at human, animal and plant genetic technologies, but most vociferously at some of the technologies of human genetics. Some of this mistrust is directed at technologies that are not yet available, or barely available, such as human genetic engineering, or even at technologies that are in fact illegal (for example, germ line gene therapy is illegal in many jurisdictions).

The technologies that I shall consider in this chapter are already in common use, so raise more practical and pressing questions. At present the new biotechnologies that have been made possible by the basic science and technologies of DNA analysis are mainly used to collect, store and interpret human genetic data, and to link them with other sorts of data.

Only a few years ago, these genetic technologies were the exclusive preserve of biomedical research and clinical genetics. Now they have many non-clinical applications. Yet they are still often viewed through the lens of medical practice. Genetic tests are discussed as if they were always done under medical authority, and were to be coupled with genetic counselling offered by medical professionals. Test results are discussed as if they were medical data, so personal to patients, and to be regulated by systems of medical confidentiality. Disclosure of genetic data to third parties – such as insurers or employers – is viewed on the model of disclosing (other) medical information.

This assimilation of genetic to medical data is becoming less and less plausible, for several reasons. First, some genetic technologies are now being put to wholly non-clinical uses, such as allocating

[1] For evidence see MORI poll commissioned by The Human Genetics Commission, *Public Attitudes to Human Genetic Information*, 2001; see institutional bibliography.

responsibility for child support payments or determining immigration status. Secondly, those who buy genetic tests (often through the Internet) can evade all medical supervision, clinical advice and genetic counselling.[2] Thirdly, genetic data about individuals can be held electronically, and genetic databases[3] can be linked, manipulated and integrated with databases containing other sorts of personal information, including non-medical information.[4] These technologies undermine the view that genetic data are a special class of medical data, and show that they may give rise to ethical issues far removed from the traditional concerns of medical ethics.

Genetic data are now used for a wide range of purposes. The longer-established medical uses include diagnosis and confirmation of diagnosis, and provision of information relevant to reproductive decisions. Incipient medical uses include large-scale epidemiological studies and other health research, deploying new and powerful developments in bioinformatics. New biotechnological uses include the use of genetically stratified cohorts of research subjects to test pharmaceuticals, in order to develop genetically targeted medicines (pharmacogenetics). Other non-medical uses range from the creation of genetic profiles or 'fingerprints' as evidence for decisions about immigration, child support responsibilities and criminal prosecutions, to the use of genetic data in setting insurance premiums or employment. In most of these areas the avoidance of deception and the maintenance of trust are likely to be among the more pressing ethical concerns.

[2] As recently as 1997 the UK Advisory Committee on Genetic Testing published a *Code of Practice and Guidance on Human Genetic Testing Services Supplied Direct to the Public* that does not engage with the reality of transnational, commercial marketing of such tests and the consequent obsolescence of such a code; see institutional bibliography.

[3] The term 'genetic database' is currently used to cover collections of tissue samples that could be used to obtain systematic genetic information about individuals, databases of interpreted DNA test results, and databases of uninterpreted DNA profiles. See House of Lords, Select Committee on Science and Technology, IIa, *Report on Human Genetic Databases: Challenges and Opportunities*, HL57, 2001; see institutional bibliography.

[4] For example, the company deCODE Genetics is linking databases containing DNA information, health information and genealogical information pertaining to the population of Iceland; see institutional bibliography.

By contrast, robust conceptions of individual autonomy that go beyond informed consent requirements are of little relevance to uses of these technologies. In the further future, some genetic technologies may enhance individual autonomy.[5] In the meantime, the ethical problems raised by current and incipient genetic technologies are more likely to have to do with the control, protection, ownership and use of genetic data, and with ways of organising these to prevent and limit deception or coercion. These problems arise out of a powerful confluence between basic DNA technologies and information technology. They are urgent and complex: genetic data can undoubtedly be used to benefit individuals and to advance research, but they can also be used in ways that may wrong or harm individuals, families and communities.

5.4 GENETIC EXCEPTIONALISM

Some commentators have suggested that genetic data are so distinctive that they should not be viewed as personal, let alone as medical, data.[6] Genetic data, they point out, are intrinsically familial rather than individual: although they must be obtained from individuals, they are not really individual data.

The phrase 'genetic exceptionalism' has been coined to express the idea that genetic data are *intrinsically* unlike other personal, including medical, data because they provide information not only about an individual from whom a sample is taken, but also about related individuals. In coming to know aspects of our own genetic make-up we may come to know something about a relative's genetic make-up. For example, if a grandparent has died of

[5] See chapter 3; see Allen Buchanan, Dan W. Brock, Norman Daniels, and Daniel Wikler, *From Chance to Choice: Genetics and Justice*, Cambridge University Press, 2000.

[6] For discussions of genetic information and genetic exceptionalism see Thomas Murray, 'Genetic Exceptionalism and Future Diaries: Is Genetic Information Different from Other Medical Information?', in Mark A. Rothstein, ed., *Genetic Secrets: Protecting Privacy and Confidentiality*, Yale University Press, 1997; Ruth Chadwick and Alison Thompson, eds., *Genetic Information Acquisition, Access, and Control*, Kluwer Academic/Plenum Publishers, 1999.

Huntington's, and the next generation have refused genetic tests, a grandchild who chooses to be tested may discover not only that he or she has inherited the gene, so will suffer the disease, but also that the parent who has refused tests has done so (knowledge is secure here, as in very few other cases, because the inheritance pattern of the Huntington's gene is dominant and the penetrance very high). A person with an identical twin may discover a genetic test result as true of her twin as of herself. More typically, people may learn from DNA test results that certain relatives may carry specific genetic variations, evidence that might (if disclosed) provide those relatives with reason, even urgent reason, to be tested or to reconsider reproductive decisions, and that might be a major source of anxiety.[7] These facts, it is said, are reasons for regarding genetic data as exceptions to the view that medical information is individual, indeed as exceptions to the view that personal data are always individual, and for revising informed consent and confidentiality requirements that assume that all personal data pertain to individuals.

Some patient support groups for genetic disorders who are attracted to the claims of 'genetic exceptionalism' have suggested that individual consent and confidentiality procedures are inadequate for dealing with collection, storage or disclosure of genetic data. They have pointed out that when members of a family suffer from a genetic condition the problem bears on all, not only on an individual, and proposed that consent procedures for genetic tests should also be familial rather than individual, and that related individuals should have a right to know information obtained from tests on any member. It is hard to see how familial consent procedures, or familial rights to the disclosure of relevant information, could be put into practice. If DNA tests required consent from (all possibly at-risk members of) families, individuals might be prevented from having a DNA test that was medically important because one or another relative did not want to learn, or want others to

[7] The effects on relatives, including children, that may be precipitated by genetic testing and 'sharing' of information range from depression and stigma to loss of marriage prospects and decisions not to reproduce.

know, potentially threatening information. Equally relatives might learn about DNA test results of which they would have preferred to remain ignorant because one relative chooses to be tested – and someone blurts out the results. And what is to happen when different relatives have different views?[8]

Indeed, there may be more than practical and emotional difficulties in the idea that genetic data are intrinsically exceptional.[9] The very notions of *genetic data* or *genetic information* are systematically unclear. A lot of genetic information is not in the first place strictly private to individuals. A great deal of loosely genetic information is evident in all our appearances, is shared promiscuously with relatives and is generally not seen as problematic. The Hapsburg lip and hundreds of other family traits are widely known and noticed, and generally taken as pleasant proof of kinship. (Occasionally such traits give rise to less pleasant queries about kinship.) In addition, some medical or personal information that is ostensibly not genetic reflects genetic factors. When insurers ask us to disclose how old our parents or grandparents were at death we do not imagine that they are interested in our family stories. We know that they want some loosely genetic information, with which we have obligingly provided them for many decades. Claims that genetic information *as such* is intrinsically special and exceptional founder on the reality that not all genetic information can be viewed as particularly private, and that much personal and medical information reflects a mix of genetic and non-genetic factors.

Even if we conclude that genetic information in general cannot be regarded as distinctive or exceptional, the more precise genetic data obtained from DNA tests might nevertheless be exceptional. Some of the information DNA tests provide can be inferred from other sources, but DNA technologies are distinctive in that samples taken from one individual can sometimes have significant and surprising implications not only for that individual but also for

[8] These issues are discussed in Genetic Interest Group, *Confidentiality Guidelines*, 1998; see institutional bibliography.

[9] Onora O'Neill, 'Informed Consent and Genetic Information', *Studies in History and Philosophy of Biological and Biomedical Sciences*, 2001a.

relatives, which nothing else could have revealed. Sometimes these discoveries are highly unwelcome. The thought that the source of a cousin's illness may be reproduced closer to home, or that an individual is a carrier for a serious disorder, is distressing: it may threaten marriage prospects, reproductive plans, insurability, future health and peace of mind.[10]

Discussions of the regulation of genetic testing have often ducked at this point, asserting not that relatives have a *right* to veto one another's DNA tests, or a *right* to be told of any DNA test results that might pertain to them, but that those who obtain DNA test results that may also be relevant to relatives should be encouraged to share the information as and when appropriate.[11] As soon as we begin to think about the many factors that might make a decision to disclose genetic test results to particular relatives either a good or a bad idea, we can see that this 'recommendation' is pretty hollow. Situations differ greatly, and so may reasonable judgements about what it is appropriate to disclose to whom in which circumstances. It is said that there is in fact a good deal of failure to disclose the results of DNA tests even to relatives for whom disclosure might be important.

Those who find the claims of genetic exceptionalism unconvincing think that refusal to disclose DNA test results to relatives raises no special ethical difficulties. They point out that many other sorts of medical conditions can be shared to some extent among relatives and more widely, but that there is no uniform obligation to share information about them. The facts of infection, contagion and of shared exposure to environmental hazards mean that cohabiting individuals – often related individuals – share a good bit of medical history: yet they have no special obligations to share information.

[10] Ruth Chadwick, Mairi Levitt and Darren Shickle, eds., *The Right to Know and the Right Not to Know*, Avebury, 1997; Ruth Chadwick, Darren Shickle, Henk ten Have and Urban Wiesing, eds., *The Ethics of Genetic Screening*, Kluwer, 1999; Chadwick and Thompson, *Genetic Information Acquisition*.

[11] Nuffield Council on Bioethics, *Genetic Screening: Ethical Issues*, 1993; see institutional bibliography. Advisory Committee on Genetic Testing, *Genetic Testing for Late Onset Disorders*, 1998; see institutional bibliography.

If we do not think the patterns created by shared non-genetic risk factors warrant departure from individual consent procedures, or from the view that medical and personal data should be treated as confidential to individuals, should we not view genetic data, too, as individual personal data, whose disclosure (if any) is a matter for individual decision?

Taken in the rather abstract way in which I have presented it, the claims of genetic exceptionalism are unconvincing. The fact that some personal data are indicative – even highly indicative – of risk for others (in this case, for related others) is matched in many areas of life. A diagnosis that is relevant for many in a family or a community may be discovered from medical data pertaining to one member. Discussions about contagion and quarantine, about sexually transmitted diseases and exposure to environmental hazards, present comparable dilemmas about disclosure to those who are at some risk but may not know it. Genetic data are not exceptional in raising these problems.

Nevertheless I believe that there are some good reasons for regarding certain uses of genetic data, rather than the data themselves, as distinctive. These data are distinctive, I believe, not so much because they are relevant to related individuals, but because third parties who have no personal interest in the data, which do not bear on their own lives or families and are not relevant to their own health, are sometimes very eager to gain access to and to use these data.

5.5 GENETIC PROFILING: UNINTERPRETED GENETIC DATA

One prominent range of uses of DNA technologies is to construct genetic *profiles* or *fingerprints*. These provide a range of structural information about (samples of) an individual's genetic make-up, but do not interpret the data. DNA *profiles* or *fingerprints* can be compared and matched without determining whether the individuals from whom the samples are taken have, or lack, any particular genetic risk factor.

For example, forensic uses of genetic information aim mainly to match the genetic profiles obtained from two samples, paradigmatically from samples taken from two different crime scenes, or from a crime scene and a suspect.[12] The procedure does not reveal which genes are included in a given individual's profile, nor therefore which risk factors are present. The process is analogous to matching uninterpreted bar codes. The profiles match or they do not; if they match (setting aside the interesting possibility of identical twins) there is an extremely high degree of probability that the two samples come from one individual. The police can discover whether there is a match without needing to find out whether a suspect is a carrier for cystic fibrosis or at higher than average risk of breast cancer. Genetic profiles provide information that can be used *in conjunction with other information* to identify persons. In the absence of other information, genetic profiles can establish only that (traces of) one and the same individual can be found at distinct locations, or that (no traces of) one and the same individual can be found at distinct locations.

So long as the information in a DNA profile remains uninterpreted, many of the issues that have preoccupied discussions of genetic testing in clinical settings do not arise. Since there will be no interpreted genetic information to impart, neither counselling nor disclosure nor non-disclosure to relatives will be an issue; nor will there be any need to think about confidentiality. The information made available by genetic profiling has no implications for individuals other than those that could have arisen from the forensic evidence with which Sherlock Holmes was familiar, such as footprints, fingerprints or fragments of fibre.[13] The difference is only that DNA evidence is more reliable.

[12] House of Lords Select Committee on Science and Technology, *Report on Human Genetic Databases: Challenges and Opportunities*, HL57, 2000; see institutional bibliography; also J. Kimmelman, 'The Promise and Perils of Criminal DNA Data Banking', *Nature Biotechnology*, 18, 2000, 695–6.

[13] A point perhaps borne out by the Home Office's amusingly titled and minimally circulated consultation document *Fingerprints, Footprints and DNA Samples*, 1999; see institutional bibliography.

Ethical concerns about genetic profiling are correspondingly likely to concentrate around relatively few issues; most of them are issues of trust. First, it is important that data are obtained only by acceptable procedures, and in particular that there is no unacceptable coercion or deception. Data are typically obtained from identifiable individuals using either clearly articulated informed consent procedures, or (where a sample is taken compulsorily) a proper exercise of public authority. If consent procedures are inadequate, or if public authority is exercised for purposes that are not essential or in ways that do not command trust, obtaining genetic profiles will be ethically suspect. Second, it is important that data are held and disclosed in ways that prevent their use for purposes that lie outside the consent given, or outside the proper procedures of the relevant public authorities. For example, if data were provided *only* for purposes of matching and identifying, any subsequent interpretation of these data, for whatever purposes, would breach the terms of that consent or the processes and authority of the relevant public body.[14]

These ethical problems and worries are neither trivial nor far-fetched. Current debates about the forensic genetic database in the UK have raised a number of questions. Should samples from suspects who are eliminated from an inquiry, or convicted only of lesser offences, be destroyed, or be destroyed after a certain period? Should police officers be required to provide DNA samples for forensic purposes, so as to make it easier to sort out which leads at a scene of crime need to be followed? Is the – generally rather popular – collection of uninterpreted DNA data for forensic purposes under strict safeguards compatible with ambitions for 'joined up government', that may require linkage of databases?[15]

[14] Certainly running paternity tests or diagnostic tests would do so. It is less clear whether research use of anonymised data would do so.

[15] Charles D. Raab, 'Electronic Confidence: Trust, Information and Public Administration', in I. Th. M. Snellen and W. B. H. J. van de Donk, eds., *Public Administration in an Information Age*, IOS Press, 1998, 113–35; Theresa Marteau and M. P. M. Richards, eds., *The Troubled Helix – Psychosocial Implications of the New Human Genetics*, Cambridge University Press, 1996; C. Weijer and E. J. Emanuel,

Will DNA data held in a forensic database remain completely isolated from other databases? May such data be used in anonymised form for research purposes? If we are not worried by these questions, it may be because we imagine that data held in a forensic DNA database can be used to the detriment only of those whose profile matches one found at a scene of serious crime, so that they will always help exonerate the innocent and convict the guilty. We have, however, only to imagine the Child Support Agency obtaining access to the forensic database to see that possibilities are wider.

DNA testing to establish paternity or non-paternity is also based on matching uninterpreted DNA test results. It raises some distinctive ethical issues. In the UK paternity tests may either be required by the Child Support Agency (when paternity and responsibility to support a child are disputed), or may be initiated by individuals. In both cases the test results pertain directly to at least two people: data from the (putative) father and from the child (or children) will be compared to see whether the match indicates paternity or non-paternity. However the test results will be highly significant for others, above all for the child's mother, but also for other (putative) siblings and more remote (putative) relatives. Yet if not prohibited, any person with access to the child and the (putative) father could obtain analysable samples of DNA without seeking consent from those from whom the samples derive, let alone from others who are indirectly affected. All that is needed are shed hairs or saliva samples. We can imagine difficult situations in which a (putative) father or a (putative) offspring commissions DNA tests without informing others whom the test results may affect, or in which a third party (perhaps a suspicious relative or neighbour) does so without the consent of any of those directly or indirectly involved. What is to stop the busybody relative who suspects that the biological father is someone outside the family, or the wrong member of the family, from testing this suspicion without the consent of those principally

'Protecting Communities in Biomedical Research', *Science*, 289, 2000, 1142–4; P. Martin and J. Kaye, 'The Use of Biological Sample Collections and Personal Medical Information in Human Genetic Research', *New Genetics and Society*, 19, 2000, 165–92.

affected, and without consideration of the interests of the child or of those who care for the child? The UK government has recently issued guidelines for providers of DNA paternity testing.[16] Given that DNA testing services can be bought via the Internet,[17] it is impossible to ensure that every paternity test commissioned from the UK, or from other states which attempt regulation, will meet prescribed standards. In particular, there is no way of guaranteeing that paternity testing will be done only with the consent of all who may be directly, and perhaps devastatingly, affected, let alone of those less directly affected.

DNA profile matching can also be used to resolve less immediate cases of disputed relationship. The possibilities are numerous and fascinating. Some are historic: the remains of the Romanovs were identified using samples given by the British royal family. Some are prehistoric: archaeologists can use DNA evidence to verify kinship and to explore historical demography. More prosaically, DNA testing is now routinely used to check claimed kinship in cases of disputed immigration entitlement. We need little imagination to see that an immigration inquiry that detects falsely claimed relationships may reveal facts unknown to some or all members of a family.

This quite brief indication of possible problems and abuses of genetic profiling suggests that DNA testing for these purposes needs tough regulation. Even if this can be provided within certain states, and even if standards are well observed where there is regulation, there is scant prospect of international acceptance of ethically robust standards. Far from pointing to a new world of enhanced individual autonomy, many applications of technologies for comparing uninterpreted genetic data need forms of regulation that restrict and control individual autonomy. The collection, storage,

[16] See Department of Health, *Genetic Paternity Testing Services – Code of Practice*, 2000; see institutional bibliography.

[17] Such tests are advertised on the Web, with emphasis on the fact that customers need not gain others' permission to have tests done. One company states on its website that since taking a hair 'is not a medical procedure, therefore one parent can collect samples without the consent of a second parent, other guardians or court'! See DNAnow.com in institutional bibliography.

disclosure and use of uninterpreted genetic data all offer opportunities for deception, and risk consequential harm. Mistrust could mushroom within states that do not create well-constructed and toughly enforced data protection systems, which provide reliable evidence that they prevent unconsented-to access and use of genetic data. Even where regulation is good, mistrust may grow if systems are opaque and the evidence they provide unconvincing or hard to interpret.

5.6 GENETIC TESTING: INTERPRETED GENETIC INFORMATION

If there are strong reasons to regulate the collection, storage, use and disclosure of uninterpreted genetic data, there are even stronger reasons to regulate the collection, storage, use and disclosure of interpreted genetic data. Genetic data must be interpreted when the aim is not just to discover whether samples match, but to discover something further about individuals from whom a sample is taken, such as a diagnosis, a susceptibility to future illness or carrier status.

Most of the discussions of genetic testing and screening that took place until the late 1990s in the UK and elsewhere assumed that genetic data would always be obtained, interpreted and safeguarded by, and to the standards of, the medical professions. DNA tests, it was thought, would be done only with a patient's consent, and only to assist diagnosis, treatment or reproductive decisions. Patients, it was assumed, could trust the medical professionals who controlled their access to DNA tests to use the new technologies for their benefit.

In order to protect patients, safeguards of varying stringency were implemented for those who were considering certain DNA tests. In particular, tests for Huntington's were offered only after patients had been extensively counselled on the implications of their results. The effectiveness of this policy was demonstrated by the fact that most who got as far as counselling (who presumably had at some stage been interested in DNA tests) declined to be tested after counselling. Tests on children were generally ruled out unless relevant to current medical treatment: testing of children for late-onset disorders was effectively barred. This essentially patient-centred view of the proper way to use interpreted DNA tests was ubiquitous in public policy and bioethical literature in the UK until the late 1990s.[18] These rather elaborate safeguards may not be feasible in a future in which more DNA information is of use in more contexts. And although tests are not yet routine in clinical practice, patients have already begun to look differently at DNA tests because they are aware that unrelated third parties, including insurers and (possibly) employers might be interested in their test results.

The long-running discussions about insurers' access to genetic test results form the most prominent debate about the claims of unrelated third parties to have access to DNA test data. These debates began in the USA, where the lack of a National Health Service obliges individuals to buy commercial health insurance, and to pay higher premiums for any factors that raise risks, including factors identified by adverse DNA test results. The UK debate, by contrast, initially centred on insurers' claims to use DNA test results as risk factors in setting premiums for life insurance. In the future questions could also arise about their use for other forms of personal insurance, including travel insurance and long-term care insurance as well as private health insurance.

The initial public inquiry into these matters in the UK was the Human Genetics Advisory Commission's 1999 report on insurance and genetics.[19] The report did not recommend legislation, or propose that insurers be forbidden to use DNA test results.

[18] Advisory Committee on Genetic Testing, *Genetic Testing for Late Onset Disorders*, 1998 and *Code of Practice and Guidance on Human Genetic Testing Services Supplied Direct to the Public*, 1997; DNAnow.com; Nuffield Council on Bioethics, *Genetic Screening: Ethical Issues*, 1993; see institutional bibliography for all of these.

[19] Human Genetics Advisory Commission, *The Implications of Genetic Testing for Insurance*, 1999 and Association of British Insurers, *Genetic Testing: ABI Code of Practice*, 1997; see institutional bibliography. See also Ruth Chadwick and Charles Ngwena, 1995, 'The Human Genome Project, Predictive Testing and Insurance Contracts: Ethical and Legal Responses', *Res Publica*, 1, 1995, 115–29; Onora O'Neill, 'Genetics, Insurance and Discrimination', *Manchester Statistical Society*, 1997b, 1–14 and 'Insurance and Genetics: The Current State of Play', *The Modern Law Review*, 61, 1998a, 716–23.

It proposed that insurers not require individuals to take tests,[20] and that they require disclosure only for test results with known actuarial implications. This proposal acknowledged the weight of the views expressed by some insurance companies, and by the insurers' trade association, the Association of British Insurers (ABI),[21] who pointed out (correctly) that insurance has always made use of medical and family history information, hence of a great deal of loosely genetic information. The ABI argued that DNA data were in principle no different from this loosely genetic information, and claimed that it would be unreasonable to forbid insurers to require disclosure of DNA test results already known to those seeking insurance. They feared that applicants who had discovered adverse information about their own life or health prospects from DNA tests would take out large policies at unfairly low rates, and that this adverse selection would harm insurance companies and ultimately other policy holders.

In fact there is little hard information on the likely costs to insurance companies of adverse selection if there were no requirement to disclose DNA test results.[22] Since insurers already have medical evidence of many of the conditions for which tests are available, and since many genetic conditions are lifelong, the *additional* actuarially significant evidence offered by DNA test results may be far less than is widely supposed. DNA tests can provide substantial new information about late-onset disorders; but where a condition is already manifest they merely confirm or improve a diagnosis without greatly altering risk predictions. Many single-gene disorders are of early onset and can be diagnosed without DNA tests. They pose no particular problem for life insurers, who can learn from medical and family histories that a condition is present. Sadly, many of

[20] On this point the Association of British Insurers agreed: although it is misleading to speak of a 'right not to know', a right to refuse to be tested is accepted by all parties in the UK debate.

[21] Association of British Insurers; see institutional bibliography.

[22] House of Commons, Select Committee on Science and Technology, *Report on Genetics and Insurance*, HC 174, 2001; see institutional bibliography. The Human Genetics Commission concurred: *The Use of Genetic Information in Insurance: Interim Recommendations*, 2001; see institutional bibliography.

those with severe early-onset conditions will never be in a position to seek life insurance.

DNA test results are, however, likely to provide significant *additional* evidence about risk levels for a small number of severe, single-gene disorders of high penetrance, which are asymptomatic until some time in adulthood, where other evidence does not disclose risk levels (even in these cases, family history evidence often discloses quite a lot about risk levels). Such conditions are quite rare, but include Huntington's, familial breast cancers (accounting only for a small percentage of all breast cancers) and familial Alzheimer's (again accounting only for a small proportion of all Alzheimer's). Even in these well-known and central cases, the actuarial implications of an adverse DNA test result are usually not known: medical validation of a test is not sufficient for actuarial validation, because actuaries need to quantify the *additional* risk of an insured-against event occurring within the period of the policy.[23]

These realities are reflected in the policy of the few UK insurance companies that have not insisted on disclosure of genetic test results, at least for life insurance policies up to a certain amount. Their approach shows that it is probably not financially catastrophic to do without disclosure of known DNA test results, at least for policies below a certain level. Nevertheless, most companies have demanded disclosure of known DNA test results.[24] I surmise that insurance companies and the ABI have taken a militant line on a supposed right to the disclosure of DNA test results mainly because of concern that the costs of any adverse selection may grow in the future if more DNA tests become available, rather than because costs would now be high.

To many members of the public, the practice of loading those with adverse DNA test results has seemed rebarbative.[25] The

[23] Harper, P. S., Lim, C. and Craufurd, D., 'Ten Years of Presymptomatic Testing for Huntington's Disease: The Experience of the UK, Huntington's Disease Prediction Consortium', *Journal of Medical Genetics* 37, 2000, 567–71.

[24] House of Commons, Select Committee on Science and Technology, *Report on Genetics and Insurance*, HC 174, 2001: see institutional bibliography.

[25] MORI and Human Genetics Commission, *Public Attitudes to Human Genetic Information*, 2001. This poll reported that only 7 per cent of the public thought that insurance

practice is not seen as analogous to that of increasing motor insurance premiums for those with high-powered cars and bad driving records, or house insurance premiums for those with thatched roofs or luxurious mansions. We choose our cars and houses: these are voluntarily incurred additional risks. We do not choose and cannot alter our genes. (Nor, of course, can we choose many of the aspects of family and medical histories whose disclosure has long been required for life and other personal insurance.)

To my ear, however, the real oddity of the Association of British Insurers' insistence on disclosure of genetic test results lies elsewhere. The traditional practice of British life insurers has been to allocate most individuals into very inclusive risk pools: in the vast majority of cases (around 95 per cent) they have taken only age, sex and smoking into account in setting premiums. Standard practice has not been to compute anything approximating an individual risk level for each applicant for life insurance. Yet the thought that, exceptional medical factors apart, any two persons of the same age, sex and smoking habit bring the same risk to the pool is evidently absurd. It is more plausible to think that this is a proper instance of risk pooling, reflecting a well-accepted degree of commercially acceptable – indeed commercially enforced – solidarity among policy holders. Even the 5 per cent of applicants who are not offered standard terms are not subjected to precise actuarial assessment, although in their case the practice has less benign results (around 4 per cent of all applicants are offered insurance at loaded rates, with loadings rising to up to 300 per cent; around 1 per cent of all applicants are refused life insurance).

One might reasonably ask whether a sudden desire for scientific rigour in calculating risks on the basis of DNA evidence fits well with the industry's strikingly broad-brush and approximate approach to other sorts of risk. One may even wonder whether it is coherent for actuarial practice to exclude racial and ethnic information from actuarial calculations (as illegal), yet insist on disclosure of DNA test results, given that a number of genetic variations

companies could be trusted to make responsible use of human genetic information in a medical database; see institutional bibliography.

are correlated with racial or ethnic background. Might this practice not count as indirect racial discrimination?

These disputes remain unresolved. The UK insurance industry agreed in 1999 that it would ask for disclosure of DNA test results only when an expert committee had established their reliability and actuarial relevance. The Genetics and Insurance Committee (GAIC)[26] was set up to advise which available tests had such reliability and relevance. This is a complex matter since actuarial evidence establishing the level of *increased* risk of a particular *insured-against event* – say, death, or disability – occurring within a certain period among those with a particular genetic characteristic may take time to collect. A genetic test may be scientifically and medically validated, yet not actuarially validated, because the *additional* risk of the insured-against event occurring within a set time is not derivable from the fact that an individual is at risk, even at high risk, across the course of a lifetime. It can be very clear that a certain test result carries a very high risk of some illness, yet quite uncertain how much the risk of illness or death is raised across the next five, ten or fifteen years. Although GAIC certified a few tests as actuarially relevant for life insurance, some UK insurers then requested disclosure of results of DNA tests not certified by GAIC, and there have been disputes about other tests. In the wake of a sharply critical House of Commons Select Committee *Report on Genetics and Insurance* in 2001,[27] that recommended a complete moratorium on all further requirements to disclose DNA test results, the industry seems far from achieving a form of self-regulation that commands public trust.

5.7 TRUST, GENETICS AND INSURANCE

I have wandered some distance into the detail and the controversy of requiring disclosure of known DNA test results to insurers. I want now to point to some deeper ethical issues that have often been glossed over.

[26] Genetics and Insurance Committee; see institutional bibliography.

[27] HC 174; see institutional bibliography.

Insurers say, and rightly, that their practices are based on trust. Policy holder and insurer are supposed to reach agreement on the basis of mutual and trustworthy disclosure of information: *uberrima fides* is the basic ethical principle governing insurance. Insurers have shown great interest in applicants' obligations to be trustworthy by disclosing accurately all factors deemed relevant to assessing their risk levels. There are sanctions for non-compliance: if policy holders are found to have deceived a company about the information available to them, they will have breached this trust, and their insurers may refuse to pay out on any claims.

However, insurers have been remarkably uninterested in the moves they need to make if their customers are to be able to reciprocate and place reasonable trust in them. Despite the pervasive rhetoric and imagery of trust in the insurance industry's advertisements – families on sunny beaches, sturdy umbrellas in storms, solid roofs on houses – it is only in the financial domain that companies are required to provide evidence of trustworthiness to their customers. This financial information is of great importance for policy holders, even if many of them find it hard to follow the small print. However, the small print does not disclose company policies on loading for risk factors of different sorts. Insurers in the UK are not required to disclose what evidence they use to load premiums: those who find themselves paying more than others are unable to establish why their premiums are higher, let alone why the loading is set at a certain level, and whether the increase is appropriate to the additional risk. They cannot even establish whether they have been asked to pay the same additional premium as other customers who presented an equivalent risk profile to the same company.

If unconditional *mutual* trust – *uberrima fides* – were really the industry's underlying principle, then it would have to view obligations of disclosure and non-deception as running both ways. It would have to achieve far higher standards of transparency and non-deception in communication, by adhering to standards that could serve as a basis for judging whether particular products were fit for purpose and represented value for money, and whether prices were set on a non-discriminatory basis. In short, the industry would have to provide the evidence and the structures required for a serious level of consumer protection.

Down the road the UK insurance industry may, I suspect, need to make a fundamental choice. Either it will have to seek scientifically well-grounded evidence about *all* risk factors, environmental and life-style as well as genetic and medical, and to introduce a serious approximation to 'actuarially fair' loading of premiums, whose known actuarial basis is then accurately and clearly disclosed to policy holders. Alternatively it may choose to continue with the rather approximate way of regarding risk information and defining risk pools that has been its traditional practice. I have no idea what the costs of the two approaches may be, although I do not imagine that across-the-board actuarial rigour comes cheap. In time the industry's policy of treating selected DNA information with far more precision than it treats other no less actuarially significant information will (I think) look incoherent if not prejudiced. With hindsight the zeal with which the industry has sought to protect its supposed right to the disclosure of DNA test results (sometimes invoking a supposed 'right to underwrite') may prove a mistaken tactic that costs insurers the very trust to which they lay claim.

CHAPTER SIX

The quest for trustworthiness

6.1 UNTRUSTWORTHY EXPERTS AND OFFICE HOLDERS

Discussions in bioethics have been marked by recurrent and deep-seated worries that experts and officials, governments and business may all be untrustworthy. In areas of concentrated specialisation and expertise, including medicine, science and biotechnology, asymmetries of power and knowledge are common: how then can inexpert patients, citizens or customers judge the experts? Can patients trust their doctors to have their best interests at heart? Can the public trust governments and their multiple agencies to regulate and fund science with proper caution and in the public interest? Should consumers trust biotech companies, whose products are developed for profit, not for public benefit? Isn't it rational to mistrust experts and officials, companies and government agencies, and to try to improve the ways in which the public can hold them to account? This litany of worries may suggest that since untrustworthy action is easy and widespread, the only safe course of action is to place no trust, thereby ensuring that no trust is misplaced.

Untrustworthy behaviour by some doctors, scientists, biotechnology companies and government agencies provides some evidence for these suspicions.[1] The most common, although not the worst, instances of misplaced trust in doctors and hospitals are the

[1] The beginnings of contemporary bioethics in the USA were animated by constant discussion of scandals, now often known simply by the names of places at which they happened: Willowbrook (the hospital that deliberately infected children with hepatitis) and Tuskeegee (where black Americans with syphilis were given no treatment in order to study the untreated disease) are among the more notorious examples.

afterglow of paternalism in and beyond medicine. It can be all too easy to calm a patient's fears with an anodyne or euphemistic description of proposed medical procedures and their effects. Patients may be told they will experience some discomfort – slight discomfort! – when a painful procedure is about to be performed: the whitest of lies, that may actually lessen pain and help patients. Prospects for recovery, or for full recovery, may be exaggerated. The dreaded word *cancer* may be avoided even if it is the key to a diagnosis. Relatives may be asked to agree to the *post mortem* removal of tissues, but not told explicitly that the tissue to be removed may include whole organs. The purposes and risks of medical experiments and drug trials on human subjects may be inadequately explained. The whole tradition of medical paternalism centred on desires to assist patients and research subjects by mild and well-intentioned deception and euphemism.

Beyond the paternalistic culture of well-intentioned deception we find notorious instances of far worse, deliberate and blatant deception of patients and their relatives. Some Nazi doctors forged mendacious death certificates for mentally ill patients and disabled people, including children, whom they had murdered. Dr Harold Shipman falsified the medical records and death certificates of the dozens of patients he had murdered. The nurse, Beverley Allitt (suffering, it is said, from 'Münchhausen's syndrome by proxy') injured children in her care and wove a web of deceit to cover up her crimes.

In these cases deception may seem a marginal wrong, because it is the lesser part of graver wrongs and crimes. But this does not mean that deception is a trivial matter: it is often indispensable in serious medical crimes, and freestanding deception can be seriously wrong in its own right. Some doctors and nurses have deceived patients to their own advantage, for example by persuading them to rewrite their wills. Sometimes diagnoses have been kept secret. Sometimes information about medical injury and incompetence has been withheld from patients, preventing them from taking remedial action or from seeking damages to which they would have been entitled. Patients have sometimes been left under the care of inadequate doctors when they could have been warned. In such cases the wrong may be 'only' deception: but it is a serious wrong.

Similar examples, running from euphemism and evasion to outright deceit and fraud, can be found in science and biotechnology, and in the public institutions that regulate them. In the UK the Phillips Report on the handling of the BSE crisis sets out various failures to be open with the public, or to act rapidly as information became available.[2] The Monsanto advertising campaign heralding the introduction of GM crops into the UK and the EU highlighted benefits, but barely mentioned risks or uncertainties.[3] In some countries, failure to inform the medical profession or the public about dangers from contaminated blood supplies (especially, but not only in the early years of the HIV/AIDS epidemic), and failure to remove contaminated blood from stock promptly, cost many lives.[4] And there are cases of outright fraud that go beyond disingenuous communication and evasion: scientists, biotech companies and journalists all sometimes misreport and exaggerate the significance of new discoveries; scientific misconduct and fraud sometimes arises from competition for grants, results and glory; peddlers of untried and untested remedies sometimes prey on desperate people. Sporadic deception can be found almost anywhere: among scientists tempted to falsify experimental data; among government agencies tempted to keep worrying medical or scientific facts confidential; among journalists tempted to exaggerate and sensationalise biomedical 'stories'; among campaigning groups eager to persuade the public of their views.

Examples of these sorts are often cited as reasons for withdrawing all trust, particularly from experts with high and impenetrable expertise, for example in medicine, science and biotechnology, and

[2] *Report of the Inquiry into BSE and variant CJD in the United Kingdom* ('The Phillips Report'), 2000; see institutional bibliography.

[3] For an account of the subsequent public debate see Parliamentary Office of Science and Technology (POST), *The 'Great GM Food Debate': A Survey of Media Coverage in the First Half of 1999*, Report 138, 2000; see institutional bibliography.

[4] Douglas Starr, *Blood: An Epic History of Medicine and Commerce*, Knopf, 1998.

from those who exercise public authority in these areas. Yet total withdrawal of trust may not even be possible. This may seem surprising. Surely trust should be withdrawn whenever there has been demonstrated untrustworthiness, and had better be withdrawn where there is inadequate evidence of trustworthiness?

Unfortunately this seemingly sensible advice is neither feasible nor coherent. It is not feasible because our lives depend in a myriad ways on medicine, science and biotechnology. We cannot avoid using them except by withdrawing from the modern world. A small number of people may choose to live the crofting life on remote islands, drink untreated water, eat only food that they grow themselves and shun modern medicine and technologies, but even they cannot control the air they breathe, or the radiation and ozone levels to which they are exposed. And for most people there is no chance whatsoever of withdrawing into self-sufficiency.

And the problem is not merely practical. The deeper difficulty is that *wholesale* mistrust is intrinsically incoherent. Those who claim to mistrust high-tech medicine, science and biotechnology *wholesale* have in practice to put their trust in something else. Some may place selective trust in alternative medicine or spiritual healing, others in the claims of 'green' campaigners or in traditional technologies. Others may place trust in religious teachings, or current fashion, or local gossip, or the suggestions of friends. Or they may place their trust in an eclectic mix of therapies, theories and technologies.

Wholesale, uncritical trust in any of these possibilities is hardly rational. Even the warmest friends of supposedly alternative, gentler, greener forms of medicine, science and biotechnology know that here too we can find well-meaning euphemism and sporadic fraud, and that the detection of deception may be no easier than elsewhere. Some partisans of the nicer, greenish remedies and fashionable therapies extol herbal medicines, aromatherapy, strenuous forms of massage and manipulation, spiritual exercises, home births and exotic diets, while offering no more than anecdotal evidence for their efficacy or even for their safety. Some claims about the dangers or the advantages of experimental 'mainstream' technologies and therapies are equally speculative. Opinions of the dangers, or of the merits of therapies, theories and technologies circulate untrammelled by evidence. Waste incineration is said to be toxic, alternatively to be preferable to disposal in landfill sites; organic crops are said to be healthier than non-organic, alternatively less healthy; tap water is said to be safer, or alternatively less safe, than bottled water; pollution is said to cause, alternatively not to cause, asthma.[5] The partisans of each of these pairs of views often go beyond their evidence, and thereby increase rather than allay doubts, anxieties and suspicions.

The proper response to incompleteness of evidence should, one would think, be more vigilant and systematic scrutiny of the available evidence and a more sustained attempt to identify which practices, therapies, theories and technologies are trustworthy, in which respects. Yet this response is uncommon: even those who trumpet their deep concern for safety are often strikingly cavalier about assessing evidence. Some make up for their selectivity about evidence by setting what they take to be commendably high standards: they insist that any risk is too much risk. Demands for unattainable levels of safety are canvassed, sometimes in the apparent conviction that certain favoured traditional technologies, or alternatively certain preferred innovations, will prove risk-free. Some of these claims may be well intentioned, if muddled; others may be deeply untrustworthy. There has always been money in selling snake oil, and there is presumably still often advantage – though hardly plausibility – in claiming that selected therapies, technologies or products are intrinsically benign and effective, even if we lack good evidence that they are safe or effective, and even when those who promote and sell them have vested interests. Some people who are unhappy about deferring to doctors are surprisingly relaxed about deferring to alternative healers and therapies.

If blanket scepticism is not a feasible basis for life we must place trust selectively and with discrimination even when we lack any guarantee that agents or institutions of any specific sort are unfailingly trustworthy. The possibility of being mistaken, deceived and even betrayed cannot be written out of life. It is therefore

[5] Richard D. North, 'Science and the Campaigners', *Economic Affairs*, 2000, 27–34.

important to find at least approximate ways of distinguishing between well-placed trust and misplaced trust.

6.2 IMPROVING TRUSTWORTHINESS

Many steps can be taken to improve the trustworthiness of practices, activities and products in medicine, science and biotechnology. Fundamental ethical obligations, the rejection of coercion and deception among them, set demanding standards. Their embodiment in legislation, regulation, public policies, institutional practice and professional standards is the first and the central way of improving trustworthiness. Good legislation, good regulation, good policies, good practices and consistent professionalism are a beginning; they need reinforcing with means of ensuring compliance, and demonstrating that compliance is reliably achieved. All this is easily stated, and hard to do. The difficulties are both philosophical and practical; I shall begin with some philosophical ones.

Ethical principles, like other practical principles, state abstract requirements. They therefore invariably underdetermine policies and action. In embedding these principles in institutions and in practice, and in choosing policies and ways of acting, agents and institutions have to find ways of resolving that indeterminacy. Some choices are made in the process of building institutions, practices or professional standards; others are made in choosing among policies and lines of action. At each stage, ethical requirements, among them requirements to reject coercion and deception in favour of trustworthiness, have to be integrated with countless other practical requirements.

Some people see a fatal theoretical difficulty in the very idea of moving from indeterminate principles to determinate structures and acts.[6] How, they ask, how can indeterminate principles ever guide action, given that every act is particular and determinate? This anxiety seems to me misplaced. It is true that we cannot

[6] The critics of 'principilism' have been particularly active in bioethics. Their concern that the specificity of cases should not be overlooked is admirable; their fear that any truck with principles will make this impossible is fortunately misplaced and arises largely out of confusion between algorithms and practical principles.

expect *any* practical principles – whether ethical or legal, social or technical – to provide a life algorithm. But the fact that principles underdetermine action means only that they must always be complemented and implemented by the exercise of judgement; and practical judgement, including ethical judgement, is not a matter of arbitrary choice. It is better thought of on the analogy of solving a design problem under multiple constraints: in this case the constraints set by the need to comply with many differing principles. In medicine, science and biotechnology, as elsewhere, ethical reasoning supplies only one of a range of constraints. Yet surprisingly, judgement is eased rather than thwarted by the commonplace fact that multiple requirements have to be satisfied. We do not generally dither about which of many possible non-deceptive acts to do, because much of the indeterminacy is resolved by the fact that we are always also pursuing a range of aims, and simultaneously meeting numerous other constraints. Practical judgement, like solving equations, may be guided rather than defeated by multiple constraints.

An example can illustrate how multiple requirements are constantly met in the commonplace task of moving from principles to practical judgement, and so to action. Anyone who aims to set up a nursing home for frail and dependent patients will have to find a building that meets the needs, that can be acquired within the budget and that complies with local planning requirements. He or she will also have to employ properly qualified staff, and to respect numerous ethical requirements such as paying adequate wages, achieving standards of care that do not risk injury or neglect of patients and refraining from fraud in obtaining loans and from abuse of the trust of frail and dependent persons. *Ethical principles are always needed in the middle of lives and activities in which action and practices, policies and institutions are constrained in multiple ways.* There is nothing mysterious about the exercise of practical judgement, or about seeking a way or ways of acting that meet a plurality of constraints. But it is often difficult.[7]

[7] Onora O'Neill, 'Practical Principles and Practical Judgement', *Hastings Center Report*, 31, 2001b, 15–23.

More generally, action that meets the ethical requirements most relevant to medicine, science and biotechnology has typically also to meet numerous requirements of other sorts. In developing a new drug, for example, scientific and clinical requirements and the legal requirements for safety and efficacy testing must all be observed, while simultaneously meeting a range of rather difficult ethical requirements, to inform, protect and obtain genuine consent from those who enter drug trials. In treating a patient with incurable disease, a care plan must be found and followed that meets an array of clinical, legal and personal requirements and constraints, and this must be done without coercing, deceiving or neglecting the patient.

Policies and decisions in medicine, science and biotechnology have constantly to meet multiple constraints, which can be variously classified as clinical, scientific, technical, legal and ethical requirements. Work in bioethics has therefore constantly, and rightly, looked for ways in which to respect and meet a large range of constraints and requirements. The non-technical and non-scientific constraints have conventionally been bundled together as 'ethical, legal and social implications' ('ELSI requirements'): but this standard and convenient grouping does not mean that ELSI requirements are ever the sole demands on action. ELSI requirements have always to be satisfied in the course of action that *also* has to meet other constraints, such as clinical, scientific, financial and technical constraints. Increasingly political constraints are also important, particularly where the structures of global free trade and of state regulatory systems make conflicting demands.[8]

6.3 THE PURSUIT OF TRUSTWORTHINESS

There is then no difficulty of principle in incorporating and living up to ethical principles in the practice of medicine, science and biotechnology: but reliable compliance has many elements

[8] For example, e-commerce makes it possible for citizens of countries in which sperm and eggs are not articles of commerce, or in which paternity testing is strictly regulated, or in which certain medicines have are not marketed because they have not been passed as safe, to buy these goods or services from less regulated jurisdictions.

and stages. Failure at any of them can damage trustworthiness. Trustworthiness sometimes fails because obligations that are important in bioethics, and rights that correspond to them, are not incorporated into legislation, regulation and the construction of institutions. Sometimes it fails despite adequate structures because there is non-compliance, reflecting either incompetence and negligence or deliberate deception and dereliction.

The first step in pursuit of greater trustworthiness is to ask how and how far structures are in place to ensure that institutions and individuals generally act in trustworthy ways. Answers will vary for different sorts of activities and in different jurisdictions. Across the last twenty-five years in the most prosperous and scientifically advanced societies, including the UK, there has been a vast amount of additional legislation, regulation and institution building aimed at the discipline and control of medicine, science and biotechnology, and specifically at ensuring that ethical standards are met. Many obligations have been imposed and allocated, and many rights secured; institutions have been constructed and strengthened; professional obligations have been clarified and codified; requirements have been placed on the public sector, on the professions and on biotechnology companies, and in particular on food processing and pharmaceutical companies. Compliance is often well checked and well enforced.

The basic aim of this legislation, regulation and control is nothing new. Many public health measures, much professional certification and many food safety measures have a far longer history. However, there is no doubt that the range and amount of legislation and regulation to control medicine, science and biotechnology have grown at a galloping pace in the last twenty years.

In the UK – parallel stories with different details might be told of a number of other countries with strong scientific and medical cultures – there have been landmark pieces of legislation such as the Animal Procedures (Scientific) Act of 1986 and the Human Embryology and Fertilisation Act of 1990. Each established a statutory body with regulatory and wider responsibilities. The Animal Procedures (Scientific) Committee regulates the use of animals in scientific research, and reports annually in great detail to Parliament. The Human Embryology and Fertilisation Authority regulates uses of reproductive technologies and also reports in detail. These are not notional or gestural pieces of legislation: each provides a tough, detailed and generally effective – if sometimes under-resourced – regime for the matters that it regulates; each has from time to time been subject to ethical criticism, but each has also earned high respect from commentators of many persuasions. Under recent and likely future legislation various previously unregulated, or minimally regulated, health practitioners – chiropodists, psychotherapists and counsellors, and 'alternative' practitioners – are being increasingly subjected to certification and regulation. Data protection legislation is increasingly used to secure the privacy of personal data in the face of electronic record keeping; freedom of information legislation improves public access to information that could previously have been kept secret. There are undoubtedly areas of medicine, science and biotechnology in which UK regulation has been or still in part is insufficiently stringent or complete: some aspects of the use of human tissues in research and some food safety issues come to mind, and in both areas a large effort to improve standards is under way. Looking across the spectrum, there are institutions that do not work as well as they should and individuals who do not live up to professional standards; improvement is possible, even if perfection is unattainable. Nevertheless there is huge evidence that an agenda of improving trustworthiness has been energetically pursued for some time.

Below the level of primary legislation there are thickets of regulatory, advisory and professional bodies that set detailed policy, enforce, review and revise regulations, conduct ethical scrutiny of research proposals, define and promulgate best practice, as well as organising public consultation and fostering critical debate about medical practice, research policy and biotechnologies. These bodies range from the Committee for the Safety of Medicines to the Gene Therapy Advisory Committee, from the Research Councils to charitable bodies with major interest in medical and scientific policy, from the ethics committees of professional bodies and Local

Research Ethics Committees (LRECs)[9] to patient interest groups and medical research charities. Many of them have worked reasonably well.

However, the regulation of bioethics was shaken, in the UK as perhaps in no other country, by the emergence of new problems in the 1990s. Although the HIV/AIDS epidemic has perhaps been less problematic in the UK than in some other countries,[10] the BSE epidemic in cattle, and the consequential (if so far limited) incidence of new variant CJD in humans has so far been largely unique to the UK, and the public response to the attempted introduction of GM crops was as vehement in the UK as anywhere in Europe. These problems were not simply responses to sporadic malpractice, incompetence or failure, which nobody would expect regulatory systems (however robust) wholly to eliminate.[11] They were signs of a more systematic crisis of public trust.

The UK government responded in 1999 by undertaking a so-called Biotechnology Framework Review.[12] On its recommendations they established three new broad, overarching bodies, each charged with taking a strategic view of a wide range of issues of bioethical concern. The Food Safety Advisory Commission, the Human Genetics Commission and the Agriculture and Environment Biotechnology Commission are each charged with oversight of a specific area, and each operates to demanding standards of openness and accessibility.[13] So although the UK has no single National Committee in Bioethics (unlike some other developed countries) it now has an unusually strong range of legislation,

[9] Department of Health, *Guidelines for Research Ethics Committees*, 2001; see institutional bibliography.

[10] See Starr, *Blood.*

[11] It is reasonable to think that detection and prosecution of some malpractice, fraud or worse is evidence that regulation is working, rather than evidence that it is failing. Total absence of detected scandals is quite likely to be evidence not of better performance but of laxer regulation.

[12] Office of Science and Technology, *Biotechnology Framework Review*, 1999; see institutional bibliography.

[13] The Human Genetics Commission, the Agriculture and Environment Biotechnology Commission and The Food Standards Agency were all established during the year following the Biotechnology Framework Review; see institutional bibliography.

of strategic and regulatory bodies, as well as strong professional structures.[14] Moreover the public culture is relatively intolerant of the merely rhetorical and declaratory products of some international bioethics declarations. There is widespread agreement that legislation and regulation are needed for ethically acceptable practice; that they must define and allocate specific obligations and rights and establish effective, adequately resourced institutions, and that diligent monitoring is needed to ensure compliance with regulation. Legislation and regulation, institutional structures and plans, as well as professional codes have often been devised with a view to securing medically and scientifically responsible and ethically acceptable control of medicine, science and biotechnology; although compliance costs are high, there is evidence of generally good compliance.

6.4 TRUSTWORTHINESS THROUGH AUDIT

Public enthusiasm for improving trustworthiness in medicine, science and biotechnology has not been deterred by incomplete success. New and more detailed measures to improve trustworthiness are constantly proposed and frequently introduced at every level. They aim to secure more trustworthy performance by enforcing more detailed compliance with more demanding prescribed procedures and practices in medicine, science and biotechnology.

These further moves to improve accountability often combine traditional and new methods. The traditional routes to accountability combined legislation and professional self-regulation. Both traditional disciplines continue. Legislation often establishes criminal and other sanctions for untrustworthy action and incentives for trustworthy action. Professional discipline and culture, whether of scientists or of health professionals, entrenches certain standards. These approaches to securing trustworthiness – or at least to

[14] Some of these may need reform to ensure that the tasks of securing professional accountability are properly separated from those of a professional guild or union; the new modes of accountability described in this section and the next sometimes undermine rather than support professional discipline and accountability.

limiting untrustworthiness – go back to the earliest civilisations of which we have any knowledge. Early legal codes sought to secure trustworthy action by punishing (for example) theft, fraud and debasement of the currency as well as by inspecting weights and measures. Early professional oaths set out the demands on all physicians.

However, in the last twenty years in many parts of the world further measures have been introduced, which introduce more precise ways of securing better and more detailed compliance with externally imposed requirements, and so (it is supposed) increased trustworthiness. A prominent feature of this widespread movement to improve accountability has been an increasing reliance on more formal procedures, including contracts, letters of agreement and financial memoranda that impose highly complex conditions. This formalisation of relations and undertakings aims to increase accountability by introducing greater clarity about expectations and about the implications of failure to meet them. It is an agenda of replacing traditional relations of trust, now grown problematic, with stronger systems for securing trustworthiness, an agenda, as John Thompson puts it, for *economising on trust*.[15] If trustworthiness can be guaranteed, then placing trust will be simultaneously risk-free and unnecessary. Formalisation has advantages that are constantly mentioned by its advocates: mutual clarity of expectations, clear performance targets, defined benchmarks of achievement, enhanced accountability.

But there is also a danger that more formalised procedures may deepen the distrust they seek to remedy. John Thompson puts the point sharply:

> in conditions of deepening distrust, legislators may be inclined to introduce more formal procedures in the hope of restoring depleted stocks of trust. Some of these procedures may indeed help, and may create greater openness and accountability of government. But there is the risk that these new procedures will only create further levels of bureaucracy and inefficiency . . . and set in motion a process that may exacerbate

[15] John B. Thompson, *Political Scandal: Power and Visibility in the Media Age*, Polity, 2000, 253–4.

rather than alleviate the problems they were intended to address, and hence contribute to a culture of deepening distrust.[16]

The very idea of 'economising on trust' is in fact ambiguous. The steps taken are designed to make trust less necessary; but the anxieties persist because trust is also less achievable.

Two powerful social agendas whose aim is to improve accountability, and so trustworthiness, may be reducing rather than increasing trust. The first agenda seeks to improve accountability by increasing and (it is asserted) improving processes of audit and monitoring, and the second by introducing greater openness into public life. I shall speak of them as *the audit agenda* and *the openness agenda*. Both innovations are likely to improve trustworthiness; but both can damage rather than restore trust.

The *audit agenda* seeks to improve accountability by ever-more intensive monitoring, inspection and audit of performance. In the UK this has been done on a vast scale during the last twenty years, mainly in the public sector and in more intensively regulated parts of the private sector. The new forms of control have been finely described and analysed by Michael Power both in his 1994 pamphlet *The Audit Explosion* and in his more recent book *The Audit Society*.[17] As Power sees it:

Audit has emerged at the boundaries between the older control structures of industrial society and the demands of a society that is increasingly conscious of its production of risks, in fields ranging from the environment, to medicine, to finance. It is one of many features of a far-reaching transition in the dominant forms of administration and control, both in government and in business.[18]

The 'new wave of audit' penetrates far beyond financial audit, although it has required institutions to conform to more specific Statements of Recommended Practice (SORPs) in financial accounting. It does not simply add to existing or traditional ways in

[16] Thompson, *Political Scandal*, 254.

[17] Michael Power, *The Audit Explosion*, Demos, 1994; *The Audit Society*, Oxford University Press, 1996.

[18] Power, *The Audit Explosion*, 6.

which the *primary* activities of institutions have been monitored, for example, by schools inspectors, University external examiners, or by the inspectors of the Royal Society for the Protection of Animals (RoSPA). Much of the new audit culture monitors the adequacy of internal *systems* by which institutions organise and control their primary activities: it is second-order auditing.[19]

Power summarises the differences between older systems of control and accountability and the new systems of the audit culture in a series of contrasts. The older systems were typically *qualitative*, often *internal* and *local*; they depended on high levels of *trust* and permitted institutions considerable *individual autonomy*; they looked at the *primary activities* of institutions in *real time*. In contrast, the new systems are *quantitative*, are *external* and often conducted at *arm's length*; they manifest *low trust* of those being called to account and exert considerable *discipline*; they look at *systems* and are typically conducted *retrospectively*.[20]

This transformation in ways of securing accountability through audit often requires changes in the structure and practices of the institutions being audited, by which they are made to conform to the new categories of audit. In effect, those who are audited are held accountable not only for achieving *outcomes* and *standards*, or alternatively for following *prescribed procedures*, but *for achieving outcomes and standards by following prescribed procedures*. Managerial and bureaucratic disciplines are to be combined in a belt-and-braces approach to securing trustworthiness. Unfortunately, the prescribed procedures sometimes obstruct rather than contribute to the outcomes and standards demanded, and sometimes distort the priorities, the aims and the efficiency of the institutions and professions to which they are applied:

> Far from being passive, audit actively constructs the contexts in which it operates. The most influential dimension of the audit explosion is the process by which environments are made auditable, structured to conform to the need to be monitored ex-post.[21]

[19] Power, *The Audit Explosion*, 6–8. [20] Power, *The Audit Explosion*, 8.

[21] Power, *The Audit Explosion*, 8; he notes also that 'audits do as much to construct definitions of quality and performance as to monitor them', p. 33; that they produce a 'drift

Power concludes that the new culture of audit achieves compliance, and thereby adds to trustworthiness, but that although it was developed to restore trust, 'its spread actually creates the very distrust it is meant to address'.[22] The new forms of audit make institutions more complex and obscure both to those who staff them and to those whom they supposedly serve. They are introduced, Power suggests, when 'accountability can no longer be sustained by informal relations of trust alone but must be formalised, made visible and subject to independent validation'.[23] The very idea of restoring trust by increased audit is doomed:

> Rather than solving the problem of trust, these models of accountability simply displace it . . . the locus of trust shifts to the experts involved in policing them, and to forms of documentary evidence . . . Ultimately there is a regress of mistrust in which the performances of auditors and inspectors are themselves subjected to audit.[24]

Despite the odium in which those subject to some of the excesses of the audit agenda generally hold it, its practices are well enforced in the public sector in the UK, well linked to financial disciplines and grudgingly accepted as supposedly effective ways of securing trustworthy compliance with externally prescribed standards and procedures. Incentives and disincentives are made clear; performance and non-performance are revealed in published league tables; independent professional judgement is often marginalised (or obstructed) by the new compliance procedures. League tables fuel competition between individuals and institutions; incentives for 'better' performance are all too plain and public. (This can create unfortunate impressions and perverse incentives: half the schools and hospitals audited will be demonstrably below average (scandalous!); distinguished teaching hospitals will have the highest

to managing by numbers', p. 34; and that 'the construction of auditable environments has necessitated record keeping demands that serve only the audit process', p. 48.

[22] Power, *The Audit Explosion*, 13. [23] Power, *The Audit Explosion*, 11.

[24] Power, *The Audit Explosion*, 13; see also Marilyn Strathern, 'From Improvement to Enhancement: An Anthropological Comment on the Audit Culture', *The Cambridge Review*, 118, 1994, 117–126.

death rates (they treat the worst cases); ratings can be more easily raised by improving process and presentation than by improving performance and standards.)

Unsurprisingly, many of those who live under the yoke of one or many of the new audit regimes have come to think that some of the forms of trustworthiness for which these regimes aim are problematic. Performance indicators have been chosen for ease of measuring, rather than because they measure relevant qualities; they often introduce perverse incentives. The audit agenda replaces serious attempts to judge quality, using care and expertise, with requirements to check compliance with procedures that may damage, or at least be oblique to, quality. Many of those who are audited find that some forms of audit and accountability damage professional standards and responsibility. Some are tempted to take a more restricted or cynical view of their professional responsibilities: those who find their clocks watched begin to watch their clocks; those who find their professional competence measured and judged by trivialising standards find that institutional loyalty and professional honour wane. It is doubtful whether the audit agenda always improves performance in medicine, science and biotechnology, even if its link to financial targets often secures compliance with prescribed procedures. Even when the new regimes of audit succeed in their own terms, they will at best produce trustworthiness rather than trust.

6.5 TRUSTWORTHINESS THROUGH OPENNESS

A second energetically pursued route to trustworthiness in recent years has aimed to construct a more open public culture, to abolish traditions of secrecy and to ensure greater publicity and transparency. The measures taken to secure openness are very different from those deployed to promote accountability and audit. Openness is supposed to achieve salutary effects on trustworthiness, and supposedly on trust, not by imposing requirements, but by ensuring that information is available to the public, including interest groups and campaigning organisations, who may then use that information to hold institutions, experts and officials to account. The expectation that one's (non-) performance will be subject to scrutiny can be expected to have a galvanising effect, and to reduce tendencies and temptations to deceive.

The openness *agenda* is well illustrated by the seven 'Nolan Principles' for the conduct of public life (see below). These principles were established by the Committee on Standards in Public Life, set up by John Major in 1994 in response to growing evidence of high and increasing public distrust of politicians and others in the UK.[25] The committee was initially chaired by Lord Nolan and published its first report in 1995.[26] Like the audit agenda, the 'Nolan Principles' address the improvement of trustworthiness rather than of trust. Their specific aim is to raise ethical standards in public life by clarifying and strengthening the ethical requirements on holders of public office. The Nolan Reports, and the wider agenda of transparency and openness of which they form part, assume but do not demonstrate that a public culture whose office holders respect the Nolan Principles, so are trustworthy, will receive commensurate public trust.

The numerous reports of the Committee on Standards in Public Life promulgate seven ethical principles for the conduct of office holders: selflessness, integrity, objectivity, accountability, openness, honesty and leadership (colloquially known as 'the Nolan Principles'). The common core of these seven principles is a demand for trustworthiness. Since the publication of the initial Reports, new and more rigorous requirements for trustworthy conduct by holders of public office have been widely implemented. Office holders are required to act only in the public interest, to be open, to avoid conflicts of interest and to declare any interest (the standard for identifying a declarable interest is that others would

[25] MORI polls reported that in 1999 that only 14 per cent of the public thought that politicians could be 'generally trusted to tell the truth', falling to 11 per cent in the case of ministers. Only journalists were more widely mistrusted; see institutional bibliography.

[26] Nolan Committee, *First Report of the Committee on Standards in Public Life*, HMSO, 1995; see institutional bibliography.

perceive it as such); declarations of interest are made public; codes of conduct and standards for public appointments procedures have been overhauled; public bodies are increasingly required to open their membership, their agenda, their recruitment process and their deliberations to public scrutiny.[27] Another example of the openness agenda can be found in the guidelines on *Scientific Advice in Policy Making* published by the UK Government's Chief Scientific Adviser. These recommend that uncertain or divided scientific advice should be made public from the start. The culture of secrecy is under pressure.[28]

6.6 INFORMATION, TESTIMONY AND PLACING TRUST

The new demands for accountability and trustworthiness that have been pursued by way of increased audit and openness, as well as by greater freedom of information, have penetrated many institutions. Their effects have been much larger than might have been imagined because new information technologies have speeded their implementation. Detailed audit and comprehensive openness need prodigious documentation. These changes depend on the personal computer and the photocopier, the fax and the website, which can provide voluminous dossiers for auditors and cheap and instant public access to information. These technologies also offer unprecedented possibilities for participation in policy debates, which earlier technologies, not to mention traditions of secrecy, did not allow. Parliamentary papers can be accessed free and without delay from any networked computer.

[27] For a useful range of evidence and selected statistics see the House of Lords, Select Committee on Science and Technology, *Report on Science and Society*, HL 38, 2000; see institutional bibliography. The appendix contains further statistical evidence of levels of trust in various professions, in which journalists consistently rank below other professions. For example, one poll on public views on meat safety found that only 0.4 per cent of respondents thought journalists the best source of information, and far greater trust was placed in scientists of whatever sort.

[28] For the 2000 version of Office of Science and Technology guidance, *Scientific Advice in Policy Making*, see institutional bibliography.

Members of the public can access information about the remit, membership, current work and reports of public bodies. Major companies, universities and charities provide increasing amounts of information on their websites. Even minor organisations and campaigning groups can make their views and proposals widely known at low cost, and often comment on and contribute to policy development. Public consultations and public hearings have become part of the repertoire of government departments, of non-departmental public bodies, of professional bodies and of charities.

Yet despite all these changes, and all these measures for improving trustworthiness, public trust still falters. Perhaps this should not surprise us. A culture of regulation, audit and openness is likely to make the institutions and individuals who must comply trustworthier: but why should it make them more trusted? Although the decades since the beginning of contemporary bioethics have seen a lot of effort to improve the trustworthiness of public institutions and of experts, culminating in the UK in the additional demands for accountability, audit and openness of the 1990s, this is quite compatible with a decline in public trust, and specifically with a decline of public trust in medicine, science and biotechnology.

One response to this seeming failure might be that restoring trust takes time, and that there has not yet been enough time for the tide to turn. In the UK some moves to increase openness and consultation took place only after the election of 1997.[29] The easier public access to public policy information requires reasonably new IT, and access is still patchy. Perhaps initial moves towards a more open public culture have accelerated exposure of previously hidden deficiencies, which are now being rooted out. Perhaps a more informed and vigilant public culture will encourage people to use the new structures to monitor and to influence, to blow whistles and to call to account in ways that improve performance.

[29] The strategic overview committees were established in 1999–2000; the intensified use of second-order audit, linked in detailed ways to financial memoranda, dates back to the 1980s and the Committee on Standards in Public Life to the early 1990s.

Perhaps increased public participation, combined with increased accountability and openness in policy processes, will not only make officials and experts more trustworthy, but will eventually revive public trust. If this optimistic story turned out to be true, efforts to improve trustworthiness would have helped restore trust. But matters may not be so simple.

An alternative and more sombre view might be that public mistrust of medicine, science and biotechnology cannot be remedied merely by ensuring that doctors, scientists and biotech companies, or those who fund and regulate them, are trustworthier. Standard political processes of reform, regulation and scrutiny cannot provide a remedy to the loss of trust *because they too are mistrusted*: how could they confer a credibility that they themselves lack? Why should anybody trust medicine, science or biotechnology just because government (who are so little trusted) have subjected doctors, scientists and biotechnology companies to arcane and opaque requirements for regulation and audit? It is notable that some of the *most* intensely regulated areas of medicine, science and biotechnology enjoy least public trust.

For a particularly vivid example of trustworthy performance coupled with public distrust we might consider the UK Animals Procedures (Scientific) Committee, whose annual reports to Parliament provide very detailed statistics on the use of animals in laboratories in the UK, the purposes for which work is done and the severity of the procedures used. The UK regulates use of animals in laboratories more tightly, to higher standards of animal welfare and is more open about the work done than any other jurisdiction. Yet in this case accountability and openness have seemingly only increased public distrust in scientists and in the pharmaceutical industry. At its most intense, distrust of scientists and biotechnology in the UK has been expressed in intimidation, criminal trespass, vandalism and even terrorism. Although a responsible minority work on and support the validation of alternative, *in vitro* methods of safety testing, most opponents of animal testing propose no serious alternative. It is sobering to realise that in this area intense efforts to improve trustworthiness have not led to any revival of public trust. Here and elsewhere it may seem that the supposed remedies are not working. Similar patterns may be noted in areas such as the introduction of GM crops, the control of gene therapy and the regulation of assisted reproductive technologies. All of these have been highly regulated in the UK at all times: yet all areas are often viewed with considerable suspicion and mistrust. By contrast, in the USA, where science and biotechnology are less regulated, trust in them is apparently greater.[30] Perhaps trust cannot *in principle* be restored by 'top-down' efforts to secure trustworthy compliance with ethically justifiable standards in medicine, science and biotechnology. Why should we expect trustworthiness, even combined with openness, to restore trust?

In the past, acceptance of others' status and authority often provided the reason for placing trust in them and in their testimony, even where evidence was scant. Those who were committed to trusting others with a certain status (friends or family, elders and priests, gentlemen and leaders)[31] felt that they had reasons for trusting the words and action of those with the relevant status. That trust might sometimes be disappointed, but was at least sometimes felt as an obligation by those whose status led others to trust them.[32] For rather good reasons we are no longer prepared to place trust on the basis of status, so have to work out whom to believe on which matters, and to judge who is competent or honest – and who is incompetent or corrupt. We can manage these tasks well in straightforward or familiar situations, and may be quite good

[30] But not trust in government, which has declined across many years in the USA, as it has in the UK. For the USA see Joseph S. Nye, Philip D. Zelikow and David C. King, eds., *Why People Don't Trust Government*, Harvard University Press, 1997; for the UK see John Curtice and Roger Jowell, 'Trust in the Political System', in Roger Jowell *et al.*, eds., *British Social Attitudes: The 14th Report*, Dartmouth 1997, 89–109.

[31] Steven Shapin, *A Social History of Truth*, Chicago University Press, 1994; Peter Lipton, 'The Epistemology of Testimony', *Studies in the History and Philosophy of Science*, 29, 1998, 1–31.

[32] Which need not be the worst of situations, as we are reminded by Samuel Johnson: 'it is happier to be sometimes cheated than not to trust', *The Rambler*, 79, 1750, 55.

at them when we have a lot of opportunity to interact with others. But placing trust is much harder where institutions are complex and remote and where interactions with others are formalised or very short term. This suggests that we may need much more than improvements in trustworthiness if we are to have any restoration of trust.

CHAPTER SEVEN

Trust and the limits of consent

7.1 THE 'CASSANDRA PROBLEM'

It was Cassandra's misfortune that her prophecies were trustworthy, but still she was not trusted. Her problem is also ours. We note only one side of the crisis of trust in medicine, science and biotechnology if we concentrate only on examples of *misplaced trust.* The other side of the problem, from which Cassandra suffered, is that of *misplaced mistrust,* unwarranted suspicion and misjudged refusal to trust, even where there is adequate – if inevitably imperfect – evidence of trustworthiness.

Cassandra's problem is neither the best, nor the worst of life's possibilities. Trust placed unsuspectingly in untrustworthy agents and institutions that deceive and betray that trust is worse. Yet mistrust directed inaccurately at trustworthy persons and institutions also leads to unnecessary anxiety (for the needlessly mistrusting) and to grief and difficulties (for the needlessly mistrusted).

Since there are never guarantees that doctors or scientists, health services or business, public officials or environmental campaigners, will always be trustworthy, there is always some risk that trust may be misdirected. Would it not then be sensible to place no trust? This approach sounds simple, but as we have seen there is no reason to think that scepticism or suspicion will always be more reasonable or safer than trust, or even that will always be feasible. Blanket scepticism may sound more sophisticated than blanket credulity, but has no more to commend it; blanket suspicion is no more reasonable or feasible than blanket trust. It is feasible to refuse to trust selected experts, specific scientific claims, ranges of pharmaceutical products, or certain alternative pundits. But it is not reasonable or feasible to refuse to trust across the board: willy-nilly we place some trust. We cannot check or monitor everything, on pain of an infinite regress.[1] So we cannot avoid the task of trying to place trust well, to trying to judge reasonably *whom* to trust on *which* issues.

In deciding how to place trust well, we can be defeated by lack of evidence, by lack of time, or by lack of expertise to assess evidence. A small sample of questions reminds us how hard it is to place trust well. Which current standards for atmospheric pollutants or food safety are too low, which about right and which absurdly high? Are there known ways of testing the safety of medicines without animal experiments? Is it safe to eat butter, or beef, or eggs, or tuna fish – or does danger lurk in them all? Is it safe to drink beer, or wine, or coffee, or fluoridated water, or tap water, or bottled water – or any water? Is it dangerous to use ecstasy, or marijuana, or is this just the message of the killjoys? What are the risks of living under power lines or telephone masts, or of using mobile phones? Are current health and safety standards or levels of medical training good enough, or are they defective? It is hardly surprising that in the face of so many complex questions, and even more opinions, trust is refused not only where there is accessible and reliable evidence of untrustworthiness, but also when there is reasonably accessible evidence for trustworthiness.

7.2 LIMITED TRUST, LIMITED SUSPICION

If we cannot live without placing trust, the much-discussed 'crisis of trust' is perhaps better described as a problem of selective mistrust, coupled with anxiety about placing trust. There is evidence enough that most people have not withdrawn *all* trust from doctors, scientists or from all pharmaceutical products, or from all public institutions that regulate medicine and biotechnology.[2] Rather they place their trust erratically and with reservations. Few of those who

[1] Michael Power, *The Audit Society: Rituals of Verification*, Oxford University Press, 1997, 2, 136; Annette Baier, 'Trust and Antitrust', *Ethics*, 96, 1986, 134.

[2] The evidence is confirmed in varying ways in numerous MORI polls; see institutional bibliography.

claim to mistrust medicine, science and biotechnology place unquestioning trust in alternative therapies, theories or technologies. How many are willing to do without intensive medical care and hospital laboratories, or to rely wholly on family doctors (and accept higher mortality rates)? How many are willing to do without modern obstetrics, and to accept the concomitant increase in maternal and infant death rates and an end to infertility treatments? How many are willing to dismantle systems for testing the safety and efficacy of medicines, and to swallow untried and untested 'cures'? How many are willing to do without pollution monitoring and control, and the sophisticated technologies on which they rely, or to trust that whatever quality of air or water we collectively produce will happen to be good enough? How many are willing to demolish the scientific quest that underlies modern medicine and so many other prized activities and products? How many in the UK are willing to give up its achievements as a centre of world-class science, generally combined with ethically demanding regulation, and to allow its scientific culture to degenerate until it fits the hoary caricature of ill-kempt boffins in garages? There are, I think, few signs that more than a small minority would welcome or pursue most of these changes.

Although some people trust some alternative medicines and therapies, and hanker for some changes in regulation, most see the alternatives as *supplements* rather than as *substitutes* for hospital-based, high-tech medicine, and are hesitant about reducing regulation. Public criticism is not, it seems, of high-tech medicine and scientific advance *as such*, nor of all regulation of medicine and science. Indeed the public have made it rather clear that they want *more and better* hospitals, *more and better* doctors and nurses, *more and better* high-tech medicine, even *more and better* science and biotechnology – and *more and better* regulation of them all.

Only when it comes to planting GM crops does the public in the UK appear to want *less* rather than *more* – a reaction that reflects public anger at the one-sided campaign of a biotech company[3] that

[3] For a neutral survey of media coverage of the Monsanto débâcle see Parliamentary Office of Science and Technology (POST), *The 'Great GM Food Debate': A Survey of Media*

trumpeted the benefits of the new technology, did not offer initial products of benefit to consumers and failed for too long to engage with public concerns about possible environmental impacts of the technology. In this area, too, the UK public has been discriminating. For example, most people buy cheese in which GM bacteria replace rennet (obtained from the stomachs of calves).

These are not the reactions of a public culture that is *intrinsically* or *in principle* hostile to advances in medicine, science and biotechnology. Rather they are the reactions of a public culture that has withdrawn trust quite selectively. Yet selective loss of trust is not a trivial matter. It can lead to free-floating anxiety, worries, uncertainties and hedging of bets and to active and corrosive mistrust directed at the trustworthy as well as the untrustworthy.

Insofar as mistrust is inaccurately directed it harms and demoralises those who are groundlessly mistrusted. It is apparent that increased demands for trustworthiness and stronger regulation of professionals and of the public sector have lead not to a restoration of trust but claims of escalating mistrust. A persistent and understandable reason for this discrepancy is that people are often not able to judge whether others are trustworthy, and that the new methods of enforcing and monitoring performance often make it no easier to do so. The sheer complexity of information and competence needed to evaluate medicine, science and biotechnology, the policies and institutions that conduct and regulate them, and the professionals who work for them, overwhelm most of us.

Yet there may be remedies for some types of loss of trust, and for some sorts of misplaced mistrust. In this chapter and chapter 8 I shall consider two quite different sorts of remedies, using as my examples medical research that uses human tissues, and public health and environmental policies. I have chosen these examples because expressions of suspicion and distrust have been particularly frequent and vehement in both areas, and because they raise quite different problems for any restoration of trust.

Coverage in the First Half of 1999, Report 138, 2000, in institutional bibliography; for the issues see Nuffield Council on Bioethics, *Genetically Modified Crops: The Ethical and Social Issues*, 1999, in institutional bibliography.

7.3 TRUST AND SUSPICION ABOUT USES OF HUMAN TISSUE

The removal, storage and use of human tissue has become an area of passionate public distrust in the UK following some well-publicised failures to seek clear consent for the retention of tissues initially removed *post mortem* under coroner's authority to determine causes of death. Such consent could and should have been sought. The reason why it should have been sought is not that informed consent provides a necessary and sufficient justification for action (as some admirers of individual autonomy imagine). Informed consent, however, is neither necessary nor sufficient for ethically justified use of human tissues: but it is generally important because it is provides an important measure of protection against coercion and deception, and also because it can make a distinctive contribution to the restoration of trust.

The ethical basis for informed consent procedures can, I have suggested, be found in principled autonomy.[4] If we are committed to principled autonomy, we must also be committed to rejecting principles of destruction, injury, coercion and deception, and to supporting others' survival and capacities to act. These abstractly stated commitments provide a scaffolding for constructing the more specific obligations and rights, institutions and policies needed in particular circumstances, including obligations to construct and maintain trustworthy institutions, practices and relationships. Without trustworthy institutions and practices, and ways of securing a reasonably good compliance with their requirements, fundamental ethical principles would too easily be violated or set aside. Informed consent requirements are one aspect of trustworthy institutions and practices. The role and limits of these requirements can be well illustrated by their part in any ethically acceptable approach to the use of human tissues.

Human tissues can be obtained in several ways, and can be used for many purposes. Most obviously, tissue that has to be removed from patients in the course of diagnosis or treatment may either be disposed of (it is then said to be 'waste' tissue) or retained, usually

[4] See chapter 4.

in the first instance for pathological examination. Such examination may be needed in further treatment of the patient from whom the tissue was taken; it may indirectly assist others' treatment. Occasionally tissue removed in the course of diagnosis or treatment may be useful for medical education, or may be referred back to in later secondary studies, thus contributing to medical knowledge, to improvements in health policies and future treatments. Sometimes the interval across which it is useful to look back is long: for example, had it not been possible to look at brain tissue retained in pathology archives in many countries across many years, it would have been hard (if not impossible) to establish that new variant CJD was a new disease.[5]

Secondly, human tissues can be removed after death. In the UK as elsewhere this may be done under coroner's authority without consent from the person who has died or of next of kin in order to determine the cause of death. In all other cases the removal of tissues can take place only with consent from the person who has died or from relatives. As a very special and tightly regulated case, tissues (and in this case frequently, but not only, whole organs) may be removed immediately after death for transplant surgery, consent having been obtained from the person who has died, from relatives, or from both. The questions that disturbed the British public in the wake of disclosures about procedures at Alder Hey Hospital, and in some other cases, arose at the intersection of coroner's authority and informed consent procedures. Nobody disputes that coroners must have authority to order *post mortem* removal of tissues to determine the cause of death. The dispute arose over the subsequent retention for research use of tissues originally removed under coroner's authority without the explicit, or sufficiently explicit, consent of relatives.

Thirdly, certain tissues may be obtained, with clear and specific consent, from individuals who are neither patients nor dead. For example, many people give blood for transfusion to others; some give sperm or eggs for use in fertility treatment for others;

[5] Personal communication, Dr Rosalind Ridley.

some give blood and other tissues for use in medical research. A few give organs and other tissues for transplanting to others, in particular to relatives. In some jurisdictions certain tissues may be sold rather than given; this is not permitted in the UK.

In all cases other than the proper exercise of coroner's authority, informed consent is important for ethically acceptable removal, storage and use of tissues. I believe that this view is correct, but misleadingly incomplete. No removal, storage or use of human tissues is rendered acceptable *merely* by the fact that the person from whom the tissue is obtained (or a relative, or anyone else) has consented. Setting aside coroners' duties to regulate the *post mortem* use of tissues to ascertain the causes of death, informed consent is always important, and in this context the consent either of the person who has died or of relatives is necessary. However, such consent does not provide sufficient ethical justification for retaining or using human tissues.

Some of the reasons why informed consent cannot offer a sufficient justification for removing, storing or using human tissues are persuasive rather than decisive. One suggestive line of thought is that if informed consent alone could justify the removal, storage and use of human tissues, there would be no reason to object to a market in human tissues. Repellent as this sounds to many people, there are such markets. In a number of countries, including the USA, individuals are permitted to sell blood,[6] sperm or eggs – for quite large sums (the *vendors* are often euphemistically referred to as *donors*).[7] In various other parts of the world organs, in particular kidneys, have been sold and bought – often illegally, sometimes with state approval. Sale and purchase are voluntary transactions, so invalid if they do not meet informed consent criteria. If informed consent were a *sufficient* ethical basis for removal, storage and use of human tissues, then either a market in human tissues would

[6] Douglas Starr, *Blood: An Epic History of Medicine and Commerce*, Knopf, 1998. Blood sales have declined in the era of HIV/AIDS.

[7] Ken Daniels, 'The Semen Providers', in Ken Daniels and Erica Haimes, eds., *Donor Insemination: International Social Science Perspectives*, Cambridge University Press, 1998, 76–104. The websites of sperm banks are also informative.

be ethically unobjectionable, or there must be some defect in all consent to sell human tissues. Since it would be hard to demonstrate that all consent to the sale of human tissues was defective we are forced to the conclusion either that a market in human tissues is ethically acceptable, or that informed consent is not sufficient ethical justification for removal, storage or use of human tissues. However, this argument is not decisive by itself: devotees of robust conceptions of individual autonomy think that a market in human tissues is indeed acceptable and that ethical requirements are fully met by securing informed consent from those who sell their tissues. Anybody who rejects this line of thought, and thinks that human tissues ought not to be bought or sold, must view informed consent as insufficient justification.

A second persuasive, but again not a decisive, reason for doubting whether informed consent is enough to justify the removal, storage and use of human tissues is that consent could be given to practices and purposes that are very widely condemned. I apologise for the beastliness of these examples: but they make the point. If informed consent were *sufficient* to justify removing, storing and using human tissues, then dismembering persons in order to provide body parts for others would be justifiable provided the heroically self-sacrificing victim gave fully informed consent to the plan. Similarly, if informed consent were *sufficient* to justify uses of human tissues, then removal of tissues that would severely damage a donor's health would be justifiable if the donor consented. Further, if informed consent were *sufficient* to justify uses of human tissues, there would be no objection to uses of human tissues that are widely seen as degrading – for example, as items of food, as ornaments or art works, or as inputs into manufacturing processes – provided that the donors gave informed consent.[8] The fact that almost nobody would regard these ways of obtaining or using human tissues as ethically acceptable suggests that informed consent requirements play their part only within a wider set of ethical

[8] Nuffield Council on Bioethics, *Human Tissue: Ethical and Legal Issues, 1995*; see institutional bibliography.

requirements that determine obligations and rights in medical and scientific practice.

7.4 THE ARGUMENTS BEHIND INFORMED CONSENT

The reason why informed consent is so much cited as the key to justifiable use of human tissues is, I believe, that we typically take for granted, but perhaps fail to attend to, a much wider range of ethical requirements that establish the purposes for which human tissues may be used and the institutional structures and professional competences needed for their ethically acceptable removal, storage and use. Human tissues, we take it for granted, are to be removed, stored and used only for medical purposes, including activities that indirectly contribute to medicine, such as medical education or medical and fundamental scientific research. We also take for granted that they are to be removed only by those professionally qualified to do so, and stored only in suitable conditions, with proper documentation and adequate safety procedures, and appropriate systems for their disposal. If we try to make informed consent requirements carry the entire burden of justifying removal and use of tissues, we overlook the basic ethical principles, obligations and rights that underlie both ethically acceptable medical practice and informed consent requirements.

The background ethical arguments for medical practice and the role of informed consent requirements within that practice can be sketched as follows. Commitment to principled autonomy – that is, commitment to principles that can be adopted by all – entails setting aside, destroying, injuring, coercing or deceiving others, and rejecting indifference to others' capacities to survive and to act. Setting aside deception entails a commitment to trustworthiness. Trustworthiness is expressed through institutions, practices and actions that it is reasonable for others to trust.

The arguments from principled autonomy do not, however, show that we have obligations never to injure, never to deceive, never to coerce, or that we have obligations to provide maximal help and maximal support to all others. Although we cannot establish maximal obligations or exceptionless rights under all these very general headings, the requirements they set are still immensely demanding. Even when agents do their best to act on principles of obligation that follow from a commitment to principled autonomy they cannot guarantee infallible trustworthiness, nor therefore can they guarantee that trust will never be misplaced.

More specifically, support for others' survival and for their capacities to act cannot be totally or maximally achieved. It is illusory to think that we could eliminate all premature death and suffering, hence illusory to think that we have obligations to do so, or that there can be a universal right not to suffer, or a right to health. But it is not unreasonable to think that rich and technically advanced societies can and ought to construct institutions that do a lot to reduce premature mortality, to compensate for many of the effects of injury and disability and to meet many basic needs, and that support for effective institutions with these aims is required of all. Any scheme for meeting these objectives will create positive, justiciable rights to medical care and will support medical progress through scientific research and the promotion of (bio)technologies that support health, and more specifically the reduction of premature mortality, suffering and incapacities. In the economically complex and globally connected societies we now inhabit, these will include commitments to the rule of law and extensive freedoms of the person, and wide-reaching regulation of the practice of medicine to prevent its covert use for unacceptable purposes. These basic ethical arguments underlie the practice of medicine and all acceptable uses of human tissues.

Given this background, we can see that it would be wrong to destroy or damage, coerce or deceive those from whom tissues are removed, and that it is obligatory to support their survival and their capacities to act. We therefore require professionals to whom removal of tissues is entrusted to be properly qualified, to use proper facilities, to follow protocols for safety and maintain proper documentation and above all to be committed to the survival and health of patients or donors, and to respect for the dead and their decisions. In short, we have strong reasons for claiming that carefully regulated practices for removing, storing and using human tissues are part of the ethically acceptable practice of medicine. Within this wider context informed consent requirements play a distinctive if subordinate role in safeguarding patients, donors and relatives, a role that can have distinctive importance in maintaining trust.

7.5 PATERNALISM AND INFORMED CONSENT IN CONTEXT

Let us imagine that we have constructed an ethically acceptable and effective health care system that is designed to prevent destruction and damage of persons, and to support their survival, health and capacities for action. We know that any health care system with these aims will inevitably concentrate a great deal of power and knowledge in the hands of its practitioners, who could therefore coerce, deceive or exercise undue influence on patients and their relatives, and also potentially on tissue donors.

Medical paternalists are unworried by this. They point out that well-directed coercion and deception by professionals can sometimes benefit patients and contribute positively to their health. Doctors may benefit vulnerable patients by being economical in giving bad news; they may benefit vacillating patients by 'persuading', even coercing, them to accept treatment. However, even if medical paternalism can benefit some patients, it is ethically unacceptable for two reasons. First, it unnecessarily gives professionals powers that can also be used to harm patients. Second, even the proper use of those powers – paternalists assume that they will only be properly used – requires and condones an unnecessary degree of coercion or deception. Failure to restrict professional powers to act paternalistically assumes falsely that these powers are needed for the proper practice of medicine, that there will be no conflicts of interest between patients and professionals and that there will be no misuse of paternalistic structures and powers.

The importance of informed consent is not simply that it protects individual autonomy – although it does so. This is fortunate, because we have already noted that individual autonomy is not always ethically important, and sometimes has to be restricted for ethically significant reasons. It follows that attempts to ground informed consent in one or another conception of individual autonomy will not reveal strong underlying arguments. It might be more revealing to reverse the direction of argument. Conceptions of individual autonomy may gain such weight as they have from their link to informed consent requirements, which are justified by their contribution to preventing and limiting deception and coercion, and in particular any deception or coercion of patients, donors and relatives.

Against this background we can accept that medical paternalism has a limited, but indispensable, role in the treatment of those who are temporarily or permanently unable to consent or refuse, but not beyond. Its justification in these contexts is not, however, that professionals have a general right or a duty to act paternalistically, but that patients without mature capacities to consent and dissent can best be protected by allowing professionals, relatives or others to act with a carefully regulated and limited degree of paternalism. Where tissues have to be removed for diagnosis or treatment, parental or proxy consent, or another institutional requirement, can provide this safeguard. Where tissues do not have to be removed for diagnosis or treatment, there is a strong case for prohibiting removal: how can those of us who are not competent to consent 'donate' blood or other tissues for research?[9] In the case of *post mortem* removal of tissues we are all in the same boat, and none of us competent to consent. Here protection is typically provided by requiring that tissues be removed only on the previously expressed wishes of the person who has died, the consent of relatives, or on the coroner's authority for strictly defined purposes only.[10]

Informed consent requirements cannot, of course, guarantee that there will be no coercion and no deception. But even if the more powerful and knowledgeable can sometimes subvert

[9] Thomas H. Murray, *The Worth of a Child*, University of California Press, 1996.

[10] The Redfern Report: *Report of the Royal Liverpool Children's Inquiry* ('The Redfern Report on events at Alder Hey Hospital') MRC, 2001; see institutional bibliography.

informed consent requirements, manipulate others' trust, or use undue influence, these requirements make coercion, manipulation and deception much harder to effect and much easier to detect. Properly used informed consent requirements can give patients who consent to surgery a veto on research use of tissues to whose surgical removal they have consented – a veto that can be exercised without cost. Properly used informed consent requirements allow relatives a veto on any *post mortem* use of tissues, other than that required to determine the cause of death. However, since informed consent requirements are not sufficient for ethical justification of uses of human tissues, these vetoes do not include powers of direction for purposes other than those mandated by the underlying ethical justification of medicine and medical research. Those who consent to removal or use of tissues do not own their tissues[11] and are not entitled to direct that they be used for ethically unacceptable purposes. Informed consent requirements need not, for example, permit tissue donors to direct that their 'gift' be sold to the highest bidder and the proceeds banked for them; or (except where donation is for a relative with compatible tissue) to direct that their gift be used for a particular individual, or for individuals with a particular social or ethnic background, or for a non-medical purpose.

The fact that informed consent procedures can offer those who consent a veto on further use of tissues means that they can contribute not only to ensuring that professionals are trustworthy, but also to restoration of trust. If these procedures are properly set up and followed, patients will be able to consent where they are willing to place their trust, and to withhold consent when they are not willing. Consent to removal of tissue for clinical reasons will not be viewed as entailing or implying consent to its use for research. The fact that a coroner orders a *post mortem* to determine the cause

[11] Libertarians dispute the point and some think that human tissues are commodities like any others. The common law does not allow for full property rights in human body parts, or for treating them as articles of commerce; nor do plausible conceptions of 'self-ownership'. See Nuffield Council on Bioethics, *Human Tissue: Ethical and Legal Issues*, 1995, see institutional bibliography; Onora O'Neill, 'Medical and Scientific Uses of Human Tissues', *Journal of Medical Ethics*, 22, 1996a, 5–7.

of death will not pre-empt relatives' decisions (or a patient's prior decisions) about any subsequent use of those tissues for research. It is reasonable to think that those who have a clear veto on research use of tissues acquire a degree of control over the use of those tissues, so gain substantial reasons to trust those who seek to use tissues for research.[12] No restoration of trust is guaranteed; but reasons to trust will be there and plainly there.

7.6 HOW MUCH INFORMATION IS NEEDED FOR INFORMED CONSENT?

In spite of the evident merits of informed consent requirements as elements in ethically adequate regulation of the collection, storage and use of human tissues, this is the very point at which problems have arisen recently in the UK. Aspects of some of the problem cases go beyond issues of informed consent. For example, the Redfern Inquiry into events at Alder Hey[13] suggests that tissues were being stored in contravention of existing regulations, in excess of any clear plans for medical research, and in unacceptable conditions, with inadequate documentation. These are clear violations of existing requirements and professional codes. But the aspect of the case that caused most distress and concern lay in a greyer area: consent to retention of tissues *after* removal to determine the cause of death had been inadequate, opaque and inadequately documented. In particular, parents who knew that tissue had been removed from their children *post mortem* complained that they had not realised that whole organs would be retained.

This episode illustrates a fundamental limit of informed consent requirements very clearly. Organs are composed of tissue, so where patients or relatives give unrestricted consent to the removal to tissue, or where relatives know that the coroner has ordered removal of tissues, professionals may have assumed that they understood

[12] The principle of giving patients greater control over the tissues they give for research can be strengthened by ensuring that they can withdraw that consent if they change their minds. This principle is prominent in the *Interim Guidelines for use of 'Surplus' Tissues* issued by the Royal College of Pathologists in June 2001; see institutional bibliography.

[13] The Redfern Report; see institutional bibliography.

that tissue may comprise whole organs. However, we have seen that consent is a propositional attitude: consent to a procedure under some description does not entail consent even to equivalent or entailed descriptions of the same act.[14] Parents were not unreasonable in insisting that although they had consented to the retention of tissue (not all had clearly done so), they had not consented to the retention of whole organs.

The initial response to publication of accounts of events at Alder Hey, and to other instances in which consent procedures were too vague, or too open to misinterpretation, or have been abused, has been a very great tightening up of consent procedures and a systematic insistence on explicitness. In my view clearer guidelines were overdue, and the sorts of issues that arose were foreseeable in outline, although not in detail.[15] We can now expect that whenever tissues are removed in the course of treatment, whenever tissue is donated, and whenever tissue is removed *post mortem*, improved consent forms will make it clear which tissues are given, and for what range of purposes they may be used.

But how detailed should the description of those purposes be? How much information is needed for informed consent to provide ethical justification? On one view we should devise highly explicit consent forms that set out in comprehensive detail which research the tissue may be used for, and even which procedures will be followed. Such explicitness about the use of tissues would raise severe practical difficulties because research purposes and secondary studies may become possible or important only long after tissues are given and archived.[16] At the time that tissue is removed it may be impossible to foresee every specific and valuable research use or

[14] See chapter 2.

[15] See n. 13. The Nuffield Council's 1995 Report did not identify all the problems that have emerged in the use of coroner's authority to permit retention of tissues, but it did identify numerous weaknesses in consent procedures for the use of 'surplus' tissues in research.

[16] For example, research to establish that new variant CJD was a new illness surveyed brain tissue archived across many years in many countries; research on ulcers was given a wholly new direction when bacteria rather than stomach acid was identified as a cause. Would it have been acceptable to delay research and treatment by insisting on *de novo* collection of tissues for either sort of research?

every significant secondary data analysis.[17] So if explicit *prior* consent is required for all research use of tissues and all subsequent data analyses, important research and important public health studies will be impossible. On some views the right way to handle this problem is to go back to patients at the later date at which new research or specific secondary studies are envisaged. This proposal too faces severe practical difficulties, not least that many patients may be untraceable or dead.[18]

More fundamentally, providing patients or relatives with complete and detailed accounts of research purposes may raise as many ethical questions as it resolves. Highly explicit consent forms that set out all possible lines of research in full detail may constitute an unacceptable burden for those who are trying to make a generous gesture at a time of stress and difficulty: an avalanche of paper, a crowd of complex questions and scores of issues to be confronted may actually diminish trust by the very ways in which it seeks to demonstrate trustworthiness. Subjecting patients to the standards of accountability of the audit agenda is ethically questionable: yet that would be the effect of using consent forms that demand explicit attention to very numerous and inevitably technically complex options. Proceeding in this direction may bring the very procedure of seeking consent into disrepute by reducing it to ticking boxes or signing paragraphs of unread fine print. Informed consent can never strictly speaking be *fully* informed: the point is rather that it should be *relevantly* informed. If it is not, consent will not be directed at the right target, and ethical justification will be lacking.

The situation of tissue donors is rather different. They are rightly given a good deal of further information about the specific research for which they give tissues, including information about risks they may run by donating. However, even donors, let alone patients and relatives, may not find that consent forms that inflict maximal

[17] House of Lords, Select Committee on Science and Technology, IIa, *Report on Human Genetic Databases: Challenges and Opportunities*, HL57, 2001; see institutional bibliography.

[18] House of Lords, Select Committee on Science and Technology, *Report on Science and Society*, HL 38, 2000; see institutional bibliography.

information are well adapted to their real capacities to provide informed consent. Donors and relatives, like patients, may find that being confronted with the full detail of research protocols provides excess, unassimilable information, to which they can hardly hope to give genuinely informed consent.

Genuine, ethically significant consent cannot, I suggest, be achieved by aspiring to some formalistic, and in principle unreachable, conception of *complete* or *exhaustive* description of the proposed research. There is a good deal of evidence from other areas of life that insisting on consent to every detail may not be the most serious or convincing way of seeking genuine consent. For example, even when we are feeling robust most of us have time and appetite for rather little fine financial print even in significant transactions like taking out mortgages or buying insurance. The difficulty of providing genuine consent to massive or complex information is likely to be even greater in the more difficult situations in which patients, donors and relatives are asked to give tissues. Insofar as the point of improving consent procedures is to restore trust, reliance on over-elaborate consent forms may be a move in the wrong direction.

In short, the demands of the audit culture and hopes of restoring trust are in tension. For audit purposes a satisfactory outcome to concerns about consent to the use of human tissues would be for every hospital to have a secure paper trail leading from every piece of tissue held to the explicitly documented, highly detailed and explicit consent of the patient, donor or relative as appropriate. Paper trails like that are ideal from the point of view of administrative quality assurance and provide good defence against possible litigation. They may be nice for hospital managers. But secure paper trails may not reassure or secure the trust of patients, donors or relatives who are asked to consent.

What might it take to secure greater trust? Those who give tissues may not be looking for a consent process in which they tick many boxes or sign off on a large number of specific propositions about the removal of tissues, that describe all the purposes for which those tissues may and may not be not used, or for the construction of a paper trail by which an auditor can check that tissues were used for the precise purposes to which consent was given. They would probably prefer a process that provides real evidence that they can choose or refuse to give tissues for research.

If changed and improved informed consent procedures are to help restore trust in medicine science and biotechnology it has to be made abundantly clear that the choice between giving and refusing lies with patients, donors or relatives. This cannot be achieved by devising maximally detailed and explicit consent forms. Genuine consent has to be based on information that is not only correct, but also accessible to patients and relatives under some stress. It must be tied to a realistic level of information about the uses to which tissues may be put. Many patients, donors and relatives might also find it easier to place their trust and to give genuine consent if they could be assured that ethical scrutiny extends to research proposals using human tissues, and that only research for serious medical and scientific purposes will be permitted. The wider context in which patients, donors and relatives can give genuine consent has to offer them good and accessible reasons for believing that they may choose or refuse to participate in a culture of solidarity, from which they and many others have benefited and will benefit.

If we seek to restore and maintain a culture of solidarity, and a better level of trust in the practices of removing, storing and using tissues, then we must aim for a *gift relationship* rather than for an audit trail. We should remember Richard Titmuss' early exploration of this theme[19] in which he noted that commercial procurement of blood could undermine the solidarity (and even the safety) that supported voluntary blood donation.[20] Extending the audit culture into an aspect of life that has until now been viewed, even if sloppily, as a matter of gift and generosity might have similar effects.

[19] Richard M. Titmuss, *The Gift Relationship: From Human Blood to Social Policy* (1970), reissued by Anne Oakley and John Ashton, eds., New Press, 1997.

[20] Institutions that prevent the trade in human tissues do not have to prohibit their transfer between medical and research institutions. But they do have to ensure that any payment is set to cover only the costs of handling and transfer. See Nuffield Council on Bioethics, *Human Tissue: Ethical and Legal Issues*, 1995; see institutional bibliography.

Ethically acceptable consent to donation of tissues should therefore, I suggest, meet three standards. First, the possibility of refusal should be made as clear and as easy to exercise as the possibility of consent. The level of information about the tissues that are given and purposes for which they are given should be clear and comprehensible rather than exhaustive or technical. Secondly, scrutiny by an independent body – an ethics committee that includes 'lay' representation – can be used to ensure that any research is for ethically warranted purposes, and that there is no use of tissues for trivial or ethically unacceptable purposes (as items for display; for research into or production of non-medical products). Thirdly, the standards of anonymisation of data should be made plain and verifiable, and subject to the same independent scrutiny. If these standards are met, consent will be *informed* in that patients, relatives and donors will know broadly which tissues and specifically which organs (described accurately, but without full scientific and physiological detail) they are asked to give. It will be *free* in that patients, relatives and donors will have good evidence that their consent counts decisively: the option of refusal will be as explicit as the option of consent, so that they have a veto on any use of tissues. (The option of refusal is often left obscure because nothing is said about what happens to tissues that are not retained. Professionals may think it obvious that tissues removed during treatment but not retained will be destroyed, or that those collected to determine the cause of death cannot always be handed back in time for a funeral: this may not be obvious to patients or relatives.)

It follows in the first place that, rather than being asked to give explicit consent to specific research projects, or to specific sorts of secondary data analysis, patients and relatives may be able to choose or refuse more confidently on the basis of more straightforward (and necessarily incomplete) information. However, in some cases patients and relatives may wish to give tissues, but have scruples about specific sorts of research. Those scruples could be respected by allowing those who give tissue to exclude its use for any research about which they have scruples. (A 'conscience' clause could allow a patient to give tissues for research, but to exclude research on, say, contraception or mental illness.) The justification for a 'conscience' clause is once again that it gives patients and relatives greater control over what happens than an ostensibly more thorough, but in fact more burdening, process of requiring consent to a full and technical description of many sorts of research.

Secondly, donors who contribute tissues for specific research projects must and can receive more information about risks. But here too a flood of complex information is not the best way to secure genuine consent, and assurance that an ethics committee that includes 'ordinary people' will have to agree that the research is important and ethically acceptable may be more convincing than reams of scientific detail.

Thirdly, patients, relatives and donors must have solid assurances about the anonymisation of data, and an undertaking that neither names nor any other identifying information will be made public. This is more than a formal requirement. Research into rare conditions and linkage studies may sometimes make it possible to identify an individual patient or donor, despite anonymisation of personal data. Patients, donors and relatives need to understand who controls which information, and in particular how any studies that link different sorts of information will be controlled.

Consent forms are not fundamental for restoring trust. Evidence that refusal is possible and respected, as well as tone, attitude and recognition of generosity are of greater importance. The generosity of those who give tissues, who allow research use of tissue that has to be removed and the generosity of relatives who give tissue *post mortem*, deserves gratitude. The best 'thank you' is not a legalistic form or an audit trail. If we merely construct more rigorous consent forms without showing gratitude where it is due we may improve trustworthiness: but we should expect no restoration of trust.

7.7 INFORMED CONSENT AND RISK

Informed consent is ethically important because it adds a tough safeguard by which individuals can protect themselves against coercion and deception. Those who have good evidence that they enjoy this protection acquire reasons to trust, since they can see that they are empowered to veto others' proposals. Since informed consent requirements can be so useful in thinking through ethical requirements for removing, storing and using human tissues, it is tempting to put them to wider work. Unfortunately they are easily overworked. Where medical, scientific or technological decisions or policies affect entire communities or groups rather than individuals, informed consent requirements often have no purchase, and can contribute little either to ethical justification or to the restoration of trust.

Many examples of loss of public trust in medicine, science and biotechnology affect groups and communities. Often they are at the intersection of environmental and public health issues, where decisions or policies affect entire communities or groups. For example, both the BSE crisis and the GM crops furore, although very different *in almost all respects*, aroused extreme expressions of public and of media distrust; and both lie at this intersection.[21] Typically environmental and public health policies do not damage, coerce or deceive identifiable individuals: if they did they would clearly be ethically unacceptable. Rather they change the risks to which individuals may be exposed. We usually cannot know whether such policies will injure any identifiable individual, but only that they may be a source of varying risk to some individuals. To be more precise, what is known beforehand is often not strictly risk information, but rather partial information about possible hazards; often there is no information about their incidence, or even about the probabilities of incidence.

No doubt, the boundaries between inflicting injury, coercion and deception and putting some others at (increased) risk are sometimes fuzzy. If a community discharges domestic sewage into a stream from which a downstream community draws its water supply they may perhaps put some others at risk. But if the upstream community is some way off, and there is good dilution, nothing may go awry until (for example) there is a severe drought. It is not obvious

[21] Parliamentary Office of Science and Technology (POST), *The 'Great GM Food Debate': A Survey of Media Coverage in the First Half of 1999*, Report 138, 2000.

how informed consent procedures could be deployed here. Yet the example is typical of many.

An initial and tempting thought, for which some devotees of the precautionary principle[22] have argued, is to think that there is a fundamental ethical requirement not to put others at risk, or at least not to put them at unconsented-to risk. However, many versions of this hurdle are unfeasible. Typically every way of living creates some risk for others, and specifically some unconsented-to risks. Unfortunately this means that strong interpretations of the precautionary principle are incoherent: they would require agents both to act in certain ways and to refrain from so acting. Traditional, unregulated farming practices inflicted many risks, for example through the sale of unsafe milk and meat, through uncontrolled infestations in food crops and above all through insecure food supply. Modernising these practices changes the risks and their incidence. High-input farming of the sort we now see in the rich world typically inflicts different risks, ranging from undetected misuse of agrochemicals, to the difficulty of making quality judgements about food that passes through many hands, from the easy transmission of pathogens among a travelling animal population to the possibly adverse effects of GM crop plants on other plants. In these cases it is common to find that *neither* maintaining the status quo *nor* any available change, *nor* any combination of changes, inflicts zero new risk, although each possible choice alters the incidence and type of risks.

When we look beyond farming and food safety to the full range of human effects on environments we find many similar cases: traditional energy sources left many unable to afford a comfortable domestic temperature, with risks to health; contemporary cheap energy policies provide more affordable domestic heating at the cost of levels of energy use that contribute to global warming, with risks for some populations (and perhaps benefits for others). Cheap energy supports affordable transport, with many benefits to standards of living, but also adds to pollution and so to ill health.

[22] See chapter 1 and below. The precautionary principle in its stronger forms proscribes action that may cause harm.

The basic problem with any attempt to use informed consent requirements to sort out which risks may be inflicted is that when the incidence of risk arising from some technology is unclear, it will be indeterminate whose consent is ethically required. Should it then be sought from all who are put at risk? It would then have to be sought from everybody, or at least from the living, who are or might be at risk, with little likelihood of agreement among them. Neither the status quo, nor any single route away from it, is likely to receive consent from all: unanimous consent to public policies is unachievable in real-world conditions.

The situation is not likely to be improved by providing those who are affected with quantifiable risk information on which to base their judgement. In the first place, it is hard to present this information in user-friendly form. Secondly, no amount of citing isolated risk figures for one option without citing those for all alternatives can get one far. Relative risk figures that compare simplified alternatives (for example, deaths per air mile, versus deaths per road mile travelled) may get one a bit further – but the definition of options is often abstract and unrealistic. Even when I realise that air travel is safer, I won't be able to make all journeys solely by air.

The problems of choosing between policies that change the incidence of risks are ubiquitous. There are no ways of living that inflict no externalities, indeed no ways of living that inflict no unconsented-to risks on others. However lightly we tread on the earth, we use resources, and above all energy. Many life-style choices – wood-burning stoves versus fossil-fuel-using central heating; organic versus mass-produced food, alternative versus scientific medicine – are choices between alternatives that both inflict risk on others, although they distribute different risks differently. Faced with these and many other difficult choices, the stronger versions of the precautionary principle have to be abandoned. More impressionistic versions that urge a certain amount of caution are unproblematic, but not especially helpful. They can no doubt offer reasons for refusing policies that clearly put many at increased risk, such as abolishing all speed limits, all food hygiene requirements or all regulation of medicines: but we do not need the precautionary principle to find these reasons.

Is the precautionary principle then no more than an illusion? Perhaps its least unconvincing use is in considering cases where one possibility clearly brings some risk of catastrophe while others do not: but many important cases do not fit this template. I am inclined to think that as soon as we face hard cases, where all options carry some risk, the stronger interpretations of the precautionary principle are incoherent, and the weaker versions lack bite. And these are the very cases for which informed consent requirements also provide little help, since it is unlikely that any particular policy will receive unanimous consent. So neither informed consent requirements nor the precautionary principle will generally help to restore public trust in policies for public health, food safety or environmental protection. Neither will offer a general remedy to the crisis of trust. Yet there may be some remedies.

CHAPTER EIGHT

Trust and communication: the media and bioethics

8.1 TRUSTWORTHINESS WITHOUT TRUST?

In chapters 6 and 7 I have looked at problems that arise when trustworthiness and trust pull apart, and at some ways in which these problems might be reduced. Being trustworthy helps in gaining trust, but is neither necessary nor sufficient. Deceivers can attract others' trust, so *misplaced trust* is common enough. The trustworthy can be denied others' trust, so *misplaced mistrust* is also common enough. Improving trustworthiness can reduce the incidence of misplaced trust; but may not attract reciprocating trust, so cannot eliminate misplaced mistrust. The 'Cassandra problem' is recalcitrant.

The problem of misplaced mistrust can, I have suggested, be resolved to a degree when trust is sought from identifiable individuals. Action can be made contingent on their consent, and forgone when they withhold consent. Consent procedures that seek genuine communication rather than legalistic form-filling, and that make it abundantly clear that dissent constitutes a veto on others' proposals for action, can help build trust. They can help to secure and maintain patients' trust in their doctors, research subjects' trust in investigators and more generally individuals' trust in others.

But this approach to restoring trust fails where action and policy bear not on identifiable individuals, but on groups or populations who are put at indeterminate risk. Only where the risks arising from some policy are unambiguously higher than available alternatives, and those arising from alternatives unambiguously less, can we regard persisting with the policy as tantamount to injuring or endangering others. The ethical principles behind fundamental health, science and environmental policies, such as those that regulate major chemical, biological and radiation hazards, or secure basic safety of air, water and food supplies, are not in doubt. Here we can reasonably view support for policies that injure or endanger as serious ethical failure. Here fundamental obligations and rights provide standards that distinguish justifiable from unjustifiable action; even dissent cannot trump these requirements. Trust may or may not be forthcoming: but obligations are clear.

But where fundamental ethical requirements cannot guide choices of health, science and environmental policies, where an appeal to basic rights and obligations is mute or indeterminate, trust in office holders, professionals, experts and regulators is both more necessary and more problematic. For reasons explored in chapter 6, doing more to ensure that they are trustworthy will not in and of itself build trust. What then should be done?

8.2 INDIVIDUAL AUTONOMY CUT DOWN TO SIZE?

One tempting thought is that nothing more should be done. Suppose that we live in a society that has successfully used legislation, regulation and other ways of securing accountability (sanctions, conditionalities, monitoring, auditing, transparency requirements, etc.) to ensure that basic rights and obligations are met. Medicine, science and biotechnology will be organised not to destroy, injure, deceive or coerce others; this will not secure universal compliance or success, but is likely to offer a reasonable level of reliability. Suppose also that we have also constructed reasonably effective institutions to support human survival and nurture human capacities; again, this will not guarantee universal compliance, but will secure a reasonable level of reliability. Fundamental obligations for the practice of medicine, science and biotechnology, and corresponding rights, would then be duly identified and reasonably well respected and enforced. Might we then have reached the limits of bioethics? Should not everything that we have no fundamental ethical reason to forbid be permitted? Once medicine, science and biotechnology are controlled to ensure that fundamental obligations are met and basic rights respected, what further case is there for control or regulation? Is not everything else the proper domain of individual autonomy?

Individual autonomy, it is widely suggested, demands a protected arena for life-style choices, especially (but not only) for those that deserve to be dignified as Millian 'experiments in living'. Individual autonomy can be expressed in choosing, for example, to eat only organically grown food, to do without a car, to recycle rubbish, to use low-energy technologies and so on. Many liberals, among them libertarians with ecological sympathies, favour these expressions of individual autonomy. However individual autonomy can also be expressed by eating cheap food grown by intensive farming, by driving a splendid and petrol-guzzling car, by joining the throw-away society, by sports and travel that use high amounts of energy and so on. Plenty of liberals, among them libertarians of consumerist outlook, favour these expressions of individual autonomy.

Since there is no automatic alliance between the pursuit of individual autonomy and either establishment or environmentalist concerns, few now place trust in life-style changes alone. Many environmentalists and other critics of mainstream medicine, science and biotechnology have overcome their original hesitation about legislation and regulation. They now doubt whether activities that bear on the environment, on public health, or on safety can responsibly be left to the play of individual autonomy or of market forces so advocate restrictive legislation and regulation. Small once seemed beautiful, but is now not thought effective enough.

Indeed, even the most ardent libertarians cannot support wholly untrammelled individual autonomy. Even where no individual act is intrinsically wrong, some public regulation of action that may create risks for others is needed for the very possibility of individual autonomy. For example, a degree of standardisation may be indispensable for any exercise of informed consent and individual autonomy. If I cannot tell whether the milk is watered, or whether the meat is locally produced, or whether the way I heat my home or travel to work is more energy-intensive than it need be, then I cannot choose in an informed way. Consumer choice itself requires enforced standards for measuring and rating products, and systematic requirements to inform consumers. Markets cannot work effectively if vendors are free to invent or distort information about products. Individual autonomy itself is undermined where options are not clearly discernible.

A second and deeper reason for doubting whether legislation and regulation can be limited to the enforcement of basic rights and obligations is that any socially and economically complex society constantly has to choose among ethically optional alternatives. By definition, this choice cannot be guided by appeal to fundamental ethical standards. For example, there may be nothing intrinsically wrong either with an agricultural system and food policy that accepts GM food crops or with one that rejects them, but (given the undisciplined habits of plants) a mixed policy might not be feasible. Neither traditional farming nor high-input agribusiness may be intrinsically wrong, but public policy still has to shape agricultural practice in one way or another. Neither fluoridation nor non-fluoridation of public water supplies lacking natural fluoride may be intrinsically wrong, but nevertheless public policy must settle for one or the other. It is simply implausible to think that beyond the measures clearly and fundamentally needed for any health, science or environmental policy that does not violate fundamental obligations everything can be a matter for individual autonomy. For all but the most enthusiastic libertarians, there will be a large gap between meeting fundamental ethical requirements and the proper domain of individual autonomy.

One tempting thought is that democratic legitimation of some sort can play a part similar to that played by informed consent requirements in legitimating action that affects individuals. There is a sense in which all public health, science and environmental policies in democracies receive democratic legitimation. They are introduced by democratically elected legislatures, or under delegated powers conferred by those legislatures, and enjoy as much democratic legitimation as any other public policy. However, many current discussions of policies affecting health, safety and the environment that are not mandated or prohibited by fundamental ethical principles suggest that they need *additional* democratic scrutiny or support.

8.3 DEMOCRATIC LEGITIMATION IN BIOETHICS

Any attempt to supplement ethical justification with (additional) democratic legitimation must recognise that democratic legitimation need not favour any specific sorts of health, science or environmental policies. Fundamental safety measures and health care provision are likely to attract a high degree of public endorsement: but these are the very policies that are in any case justified as required by fundamental obligations and fundamental rights; their claims would not lapse even if democratic endorsement failed. By contrast, policies that curb consumer choice or require higher taxes for ethically optional health and environmental purposes often attract only limited public support, so may receive neither ethical justification nor political legitimation.

Democratic legitimation is ethically unreliable. It may not endorse action that is ethically required, or action that is important for minorities, or action that protects the environment, public health or public safety; and it may endorse action that is ethically unacceptable and risky in many ways. An interesting example of these problems in bioethics arose out of a well-known public consultation in the State of Oregon, which was intended to legitimate policies for rationing publicly funded health care. Many members of the public gave low priority to funding treatment of certain severe but unpopular conditions, such as mental illness and HIV/AIDS, which they may have associated with life-styles they condemned.[1]

[1] See Martin Strasbourg, Joshua Wiener and Robert Baker, with I. Alan Fein, eds., *Rationing America's Medical Care: The Oregon Plan and Beyond*, Brookings Institution, 1992. For a reassessment of what actually happened in Oregon, see Lawrence Jacobs,

Here democratic legitimation apparently conflicted with basic equity in health care provision.

Democratic legitimation also constantly conflicts with other ethically significant principles and goals. Democratic processes need not endorse passionately held green views: people like cheap food, despite its high environmental and animal welfare costs. They need not endorse green conclusions: people like their cars. They need not endorse healthy conclusions: people like their chips and their cigarettes. There is no necessary link between democratically legitimated policies and ethical requirements, green positions or public health. It may nevertheless be true that democratic legitimation can increase public trust.

How could it do so? One thought is that democratic legitimation can parallel the contribution informed consent makes to building trust where action affects individuals. Perhaps extra public discussion, consent and endorsement can legitimate ethically optional policies, just as individual informed consent can legitimate decisions affecting individuals. The public will then not be able to dismiss a policy merely as advocated by government, or by experts, or by campaigning groups, or by others whom they may not trust. But the two cases are not parallel. Individual consent procedures ensure that nobody is outvoted or overruled; procedures for democratic legitimation invariably leave minorities outvoted or overruled, if able to express dissent. Those minorities may continue to see policies endorsed by the majority as unacceptable, and those who devise, advocate or implement them as misguided and untrustworthy.

A brief consideration of some current proposals for securing enhanced democratic legitimation throws up many further problems. I list some possibilities and some shortcomings at high speed: neither list is complete. Some of the proposals are for (better, increased) consultation of the public without additional public deliberation; others canvas increased or new forms of public deliberation.

Theodore Marmor and Jonathan Oberlander, 'The Oregon Health Plan and the Political Paradox of Rationing: What Advocates and Critics Have Claimed and What Oregon Did', *Journal of Health Politics, Policy and Law*, 24 (1), 1999, 161–80.

One long-standing form of public involvement uses *opinion polls* and *consultations* to record a snapshot of public views on a range of issues at a given moment. The results can be valuable in making policy, even if there is no actual deliberation, even if questions have to be simplified and even if only limited information can be provided. For example, the Human Genetics Commission conducted an informative public consultation in 2001 that asked for public views about what *could* and what *should* be done using genetic information. The consultation revealed, for example, that a very large majority knew that insurers *could* use genetic data, but that very few, and even fewer of those who were better informed about genetics, thought that they *should* do so: this is surely valuable information about public understanding and public views[2]: but it does more to inform than to legitimate policy.

Another type of public consultation addresses only those most directly involved, or the organisations most immediately affected, and takes account of their views in forming policy. This is widely, if sometimes simplistically, done; even when it contributes valuably to policy-making it may contribute little democratic legitimation. For example, in the summer of 1999 the Home Office consulted a limited range of professional and official respondents on a range of topics grouped under the intriguing title 'Fingerprints, Footprints and DNA Samples'.[3] Since then the national criminal genetic database has grown apace[4]: but any claim that it received a degree of democratic legitimation from this selective and lightning consultation must be extremely slender. So far the use of genetic information for forensic purposes has in fact proved popular, presumably because it is thought to increase convictions for serious

[2] See references in chapter 5 and institutional bibliography. The Human Genetics Commission survey of public attitudes to genetics in 2001 showed that four out of five persons thought that insurers should not use genetic data to set premiums. Exact figures differ with the type of insurance.

[3] Home Office, 'Fingerprints, Footprints and DNA Samples', 1999; see institutional bibliography.

[4] House of Lords, Select Committee on Science and Technology, IIa, *Report on Human Genetic Databases: Challenges and Opportunities*, HL 57, 2001; see institutional bibliography. See also the published evidence, HL 115, 2000; see institutional bibliography.

crimes. It may be less popular once a higher proportion of the population have had samples taken and retained, particularly if it were accepted that samples from persons not later convicted should be retained.[5]

Even when consultations elicit a wider sample of views, legitimation may be limited. In general, only a small proportion of the public respond to consultations; the views of establishment and campaigning organisations may be disproportionately visible in the responses; subsequent policy-making is of necessity in the hands of officials and experts. Consultation is surely better than no consultation, but adds only minimal democratic legitimation.

A more direct and perhaps more promising approach to democratic legitimation uses the device of a *citizens' jury* to obtain an integrated and reflected sample of public views. Citizens' juries bring together small groups of citizens who supposedly represent the general public. These individuals

> meet to deliberate upon a policy question . . . [and] are informed about the issues, taking evidence from witnesses and cross examining them . . . [and] then discuss the matter among themselves and reach a decision.[6]

When well conducted, citizens' juries can achieve an enlightening exchange of views. But there are many pitfalls between that exchange and any democratic legitimation of its outcomes. Some citizens' juries reach only anodyne conclusions because they have not been organised to provide focused answers.[7] Others do not contribute much to democratic legitimation because some interest group, such as a business or trade association, or a lobbying or campaigning organisation, funds the occasion, frames the questions, selects the 'jurors', controls the information they receive and then

5 For up-to-date information on the UK Forensic Database see institutional bibliography; see also chapter 5.

6 John Stewart, Elizabeth Kendall and Anna Coote, *Citizens' Juries*, IPPR, 1994, iii.

7 *Report of the Citizens' Jury on Genetic Testing for Common Disorders*, Welsh Institute for Health and Social Care, University of Glamorgan, Pontypridd, 1998, which addressed the question 'What conditions should be fulfilled before genetic testing for susceptibility to common diseases becomes widely available on the NHS?'; see institutional bibliography.

disseminates (some of!) the 'findings'. Procedures allegedly undertaken to consult the public, or to improve public understanding, or to increase public involvement and participation, and so ultimately to legitimate the outcomes of a policy process, are readily subverted to magnify the public relations agenda of one institution or group or another. Any manipulative use of consultative and deliberative processes undermines and does not help build public trust; indeed it may undermine public trust in public consultation and citizens' juries. Even when there is no hidden agenda, and no evident manipulation, citizens' juries are an expensive way of enabling a small sample of the public to gain information and deliberate. They contribute more to the engagement of that smallish cohort with health, science and environmental or other public policy than they do to democratic legitimation.

The same may be said of somewhat larger-scale *citizens' forums* and *consensus conferences*. Even those who have been members of the juries, or panels, or conferences may see little reason to accept their recommendations, or to take them into account in placing trust. Those who have taken no part may feel (sometimes with good reason) that those who did were co-opted by the very office holders, experts, industries and campaigners whom they have difficulty in trusting. They are likely to regard the whole process as an adjunct to somebody's public relations agenda; and they are not always wrong.

Finally it is imaginable that more traditional forms of democratic legitimation could be achieved by increased use of referendums, which reach all citizens but demand no participation or deliberation. Here the difficulties are different. Since an actual debate involving all members of the public is not feasible, referendums – even when the question put is carefully controlled, and information carefully provided – may yield answers that are incompatible with other required or favoured policies, or with respect for human rights or treaty obligations, or with existing laws or financial resources. Yet if questions are too carefully controlled, and the agenda too tightly set, the degree of democratic legitimation may be rather slight.

Each of these approaches to democratic legitimation might add to public trust if well handled. But they can unfortunately also contribute to public distrust, for example because many members of the public will not take part, and may see no reason to trust those who did take part. Meanwhile policy-makers and professionals, companies and campaigning organisations, will always be wary of assigning any systematic role to the outcomes of these processes. They will note the basic shortcomings of supposed expressions of public opinion conducted under such limiting conditions. They will find little analogy between careful processes of seeking informed consent from individuals who are able to refuse participation, and the processes of political legitimation to which these consultative and deliberative exercises aspire.

These rather disappointing features of various current approaches to democratic legitimation all arise either from absence of sufficiently well-informed debate, from absence of a tough focus on real options, or from absence of representation, or from a combination of all these absences. It is not likely that any of these supposed modes of political legitimation would restore trust. From the point of view of most members of the public, the supposed processes of legitimation simply raise the question whether to trust the conclusions of others who were consulted or who deliberated.[8]

8.4 BIOETHICS AND THE MEDIA

An alternative approach to building public trust in medicine, science and biotechnology might begin with the thought that we in fact have a robust and widely accessible forum in which they are constantly discussed. Both print and broadcast media conduct a constant, complex and multi-faceted debate on many of the issues of bioethics. Cannot this debate contribute to restoring public trust, and perhaps also to public policy formation? Indeed, might not the media debate supply or contribute to some form of democratic legitimation?

[8] For some limitations of deliberative democratic procedures see Robert Goodin, 'Democratic Deliberation Within', *Philosophy and Public Affairs*, 29, 2000, 81–109.

A fundamental reason for doubting whether this debate, as presently conducted, can contribute to restoring public trust is that the public also does not trust the media. Reported levels of trust in newspaper journalists are generally far lower than levels of trust in all holders of public office, all professionals and in all scientists and experts of all sorts – indeed, even lower than levels of trust in politicians.[9] Why, then, should the public trust what those journalists report? (Why, specifically, should they trust them to report who is more and who is less trusted?) Although reported levels of trust in TV commentators are higher (generally higher than levels of trust in politicians), the thought that the media could help restore trust is not immediately plausible. In this case, it may be thought, the public have good reason not to trust. Unlike politicians and civil servants, and those who work in the (mainly) public sector institutions that practice and regulate medicine, science and biotechnology, the print media are subject to almost no regulation or audit (beyond financial audit), and TV journalists to rather little.

As we have seen, although there are no guarantees there are quite strong reasons for expecting doctors or scientists or civil servants to be trustworthy: they face large penalties when they are not, and misconduct is reasonably likely to be detected. Untrusted politicians can at least be voted out of office. By contrast newspaper journalists face few disciplines that support public trust. They must get 'stories' accepted by the editor (a revealing change of idiom: time was when good journalists eschewed *stories*); they must avoid libelling the rich and famous (libelling the less-than-rich is generally risk free, since there is no legal aid for libel). And there

[9] For a useful range of evidence and selected statistics see the House of Lords, Select Committee on Science and Technology, *Report on Science and Society*, HL 38, 2000; see institutional bibliography. The appendix contains ample statistical evidence of the level of trust in various professions, in which newspaper journalists consistently rank very much below other professions. For example, in a 1999 poll on views on food safety only 0.4 per cent of respondents thought journalists writing for a newspaper would be the best source of information, but 26.7 per cent reported that they would place most trust in scientists working . . . for the meat industry! (Trust in scientists with other employers was in some cases higher and in others lower: but always dramatically higher than trust in newspaper journalists.)

are some not-too-demanding restrictions on publishing obscenity. Of course, employers may impose further disciplines, and these can weigh heavily on conscientious journalists: but they may or may not offer reasons for the public to place trust in newspaper articles. For example, journalists may be told what line to take in a 'story', or may find their employment at risk if too few of their 'stories' are published. Neither editorial requirement will give the public reasons for placing trust.

For newspaper journalists there are no enforceable requirements for accuracy or coverage or balance; there are no enforceable requirements to refrain from writing on subjects of which they are ignorant; there are no enforceable requirements to distinguish reporting from commentary. Moreover in place of requirements to acknowledge sources there is a well-guarded 'right' to hide sources, that can be used to obstruct the reader's ability to tell whether there is any source whatsoever, or (if there is) whether it can trusted. Public mistrust of newspaper journalists is hardly surprising, and galling for those of them who try to maintain reasonable standards.

In the UK, and also in the USA, broadcast journalism is regulated to a much greater extent. In the UK the Broadcasting Act 1996 requires the Broadcasting Standards Commission to publish a code applying to all broadcasters and to consider complaints. Within this regulatory framework broadcasters can set themselves more explicit commitments. The BBC, for example, is precluded by its Charter from broadcasting its own opinions on current affairs or matters of public policy. Its *Producers' Guidelines*[10] set out additional commitments to editorial values that include *impartiality*, *accuracy*, *fairness*, *giving a full view*, *editorial independence*, *respect for privacy*, *standards of taste and decency* and *safeguarding children's welfare*. These standards are less demanding than those that apply in the professions or in the public sector. Broadcasters are not, for example, required to declare their interests, or even their conflicts of interest, or to withdraw from broadcasting on topics in which they

[10] British Broadcasting Corporation (BBC), *Producers' Guidelines: The BBC's Values and Standards*, BBC, 2000.

have a financial interest. Nor does broadcasting practice, even in the BBC, always meet these standards.

The effects of this state of affairs on reporting on medicine, science and biotechnology are not surprising. Some reporting fails in accuracy, impartiality and fairness, and quite often in respect for privacy, standards of taste and decency (according to some critics also in safeguarding children's welfare). These are not just the jaundiced impressions of those who prefer more academic standards. Comparative studies of the British press show that it has become more partisan and more dominated by scandal and sensationalism than newspapers elsewhere, and that these tendencies spread from tabloids to 'quality' papers during the 1990s. Pieces of accurate and careful reporting and of penetrating analysis can be found in many papers (more in some than in others), but few even of the 'quality' papers take steps to ensure that their reporting is systematically accurate or careful.[11]

This may not matter for highly sophisticated readers who have other sources of information that enable them to distinguish 'stories' that are near to the mark from those that are false or confused, vicious or silly. Many readers are not in this position. For example, press coverage of issues on GM crops in 1999 included responsible articles, but few readers without a reasonable knowledge of the science and access to quite extensive independent information could have hoped to distinguish tolerably accurate information, from misinformation, or from disinformation. A cautious and neutral survey of media coverage of GM crop issues in this period[12] concluded that some newspapers were ' "fuelling" the debate and setting the agenda',[13] in short acting as campaigning organisations:

[11] Holli A. Semetko, 'Great Britain: The End of News at Ten and the Changing News Environment', in Richard Gunther and Anthony Mugham, eds., *Democracy and the Media: A Comparative Perspective*, Cambridge University Press, 2000; Shirley Dex and Elaine Sheppard, *Perceptions of Quality in Television Production*, Judge Institute for Management Studies, Cambridge, 2000; John Thompson, *Political Scandal: Power and Visibility in the Media Age*, Polity, 2000.

[12] Parliamentary Office of Science and Technology (POST), *The 'Great GM Food Debate': A Survey of Media Coverage in the First Half of 1999*, Report 138, 2000.

[13] POST, The 'Great GM Food Debate', 41.

they had given up reporting accurately. Even the broadcast media did not maintain impartiality, as evidenced by a notorious interview of the then Minister of Agriculture, Dr Cunningham, by the well-known broadcaster John Humphries, interviewing for the influential *Today Programme*.[14]

These problems are not, of course, confined to newspaper coverage of bioethics. They are pervasive. In his moderately vituperative *The New Elites*,[15] George Walden describes the British media as having assumed the characteristics of a traditional elite. They are cavalier both towards democratically constituted authority and about individual rights of privacy;[16] they are impatient of restrictions on their activities; they pose as the voice of the people and the arbiter of morality; they remain untouched by scandal mainly because 'their discretion about the peccadilloes of individual [journalists] is as complete as their condemnation of the same failings in others is unhesitating'.[17] Walden notes that 'as they ask probing questions about jobs for the boys in politics, the media itself remains a bastion of patronage',[18] and berates a world in which journalists purport to speak in the name of the people yet show contempt for the institutions of democracy and the views of ordinary people. He heaps scorn on 'fortunate, well-schooled folk [who] set about making a career in the masses and [show no] squeamishness when it comes to exploiting the tastes of people less well educated than themselves in the furtherance of those careers'.[19]

These are criticisms of the broadcast as well as the print media, and many are apt. While broadcasters are not allowed to be politically partisan, they too commonly give airtime to 'stories' that have little foundation or importance, and use a range of

[14] POST, The 'Great GM Food Debate', Appendix D, 49–52. Humphries is an organic farmer, and publishes on organic farming: see John Humphries, *The Great Food Gamble*, Hodder & Stoughton, 2001. He was not required to and did not declare his interest; nor was he required to hand over reporting to colleagues without any financial interest in the topic.

[15] George Walden, *The New Elites: Making a Career in the Masses*, Allen Lane, The Penguin Press, 2000.

[16] Walden, *The New Elites*, 173.

[17] Walden, *The New Elites*, 173.

[18] Walden, *The New Elites*, 175.

[19] Walden, *The New Elites*, 181.

techniques that are likely to distort. Interviewers frequently ask leading questions and put words into others' mouths; some even describe themselves as 'tribunes of the people' – although they lack a democratic mandate for their unaccountable position and power. Some programme-makers treat newspaper innuendo as news in its own right, and recirculate the worst of tabloid 'journalism'.[20] Chat show 'hosts' encourage their 'guests' to disseminate gossip and disinformation in their bids to draw attention to themselves. Accurate reporting of events of public interest, scandalous or not, is often subordinated to gossip about supposed scandals of little public interest.

One does not have to share all of Walden's fury with a media culture that so often replaces reporting with gossip[21] and scandal mongering to see that reasons for trusting many parts of the British media have declined. Once too many 'stories' are cavalier about basic requirements for informative and accurate reporting, the public political culture is damaged. Bad reporting drives out good reporting, just as bad money drives out good money.

Reporting on medicine, science and biotechnology is particularly vulnerable to this sort of inadequacy. All too many topics in these areas have aspects that are technical and boring, and many are incomprehensible without coverage of arcane and complex evidence. Yet some of these topics lend themselves to sentimentalising and sensationalising ways of writing. Reporting on medicine, science and biotechnology does not generally attract either the prestige or the quality of writing that can be found in the best reporting of political and economic news. Cynicism and mistrust are understandable – indeed reasonable – responses to a media culture that freely mixes reporting with sensationalism, sentimentality and misrepresentation.

[20] For discussion of TV quality, see Dex and Sheppard, *Perceptions of Quality in Television Production*.

[21] The complaint is not new: 'Gossip is no longer the resource of the idle and of the vicious, but has become a trade, which is pursued with industry as well as effrontery', Samuel D. Warren and Louis D. Brandeis, 'The Right to Privacy', *Harvard Law Review*, IV, 1890, 194–219.

8.5 PRESS FREEDOM AND BIOETHICS

Many sober people in the UK now hold these or similar discouraged and discouraging views of the newspapers, and (to a large degree) of broadcasting. In particular they may find a lot of media coverage of medicine, science and biotechnology intrusive and offensive in tone. They may be despondent about sentimental and intrusive 'stories' about individual patients, and about sensationalising coverage of health and environmental questions that matter deeply to them. We have only to think of the crude inaccuracies of some reporting on 'genes for' social characteristics, on genetic determinism, on GM crops, or on imminent miracle cures to think that the media could do much better. In reflective moments these critics of the media often point to the pressures arising from innovations in communications technology and the pressures of global competition and the ways in which these are eroding the standards, perhaps even the possibility, of public service broadcasting as it arose and flourished in the twentieth century.

Yet on rehearsing these gloomy thoughts many people shrug their shoulders, because they believe that the price of having a free press includes having to put up with shoddy reporting. They conclude that, even if the present habits of large parts of the British media are deplorable, even if the culture of leer and smear and jeer, of name and shame and blame, is sickening, still there is no available remedy other than censorship – which is unthinkable. They conclude that there is no acceptable way to have journalism that is both honest and critical, and that we must put up with reporting, including reporting on medicine, science and biotechnology, that is unaccountable and without standards.

I have come to think that this reaction, often my own, is unconvincing. Broadcasting is considerably regulated, but we do not view this regulation as censorship. Freedom of the press does not have to be unconditional freedom, and it does not require abysmal standards. A reasonable level of accuracy in reporting is not even particularly difficult: broadcasters and newspapers achieve it quite routinely in reporting sports results and stock prices. Broadcasters and many newspapers have the capacity to report fully and critically on complex affairs when it matters, for example in covering a serious political crisis. They could extend the same capacities to coverage of bioethics. A very small number of newspapers, and some bits of other newspapers, do so, and maintain reasonable standards in their reporting on medicine, science and biotechnology.

These are issues of such fundamental importance, in bioethics and beyond, that it is worth considering the basic obligations and rights that govern all communication, including mediated communication. I begin with a brief reminder of the powerful lines of argument for freedom of the press that we inherited from the nineteenth century. Somewhat to my surprise, they neither individually nor jointly show that it is wrong to regulate the media *when it is necessary for better communication.*

First, it has traditionally been argued that freedom of the press is necessary for the emergence of truth. An appeal to this argument by large parts of the British press would currently sound like a bad joke, given how many newspapers and journalists are cavalier about reporting only the truth. The noble but archaic image of the journalist as one who speaks truth to power is obsolete in a world in which the media are themselves among the least accountable powers, and can choose to publish material that they know to be false as well as material that they do not know to be true. Even if the media are not powerful in the ways in which armies and governments can be powerful, still they have and use their power to make and break governments, companies and individuals. The power to set the agenda, to make and break reputations, especially the reputations of the glamorous or powerful, is a mighty power; it is often deployed with snickering glee. In a democracy anybody, and any institution, that has such powers may rightly be held accountable for their use.

The argument that the media must be free because this serves the pursuit of truth might seem less bizarre if we imagined that they do so collectively, by enacting some version of conjecture and refutation. However, it seems very unlikely that the sorts of reporting of medicine, science and biotechnology that flourish at present are optimal for, or even support, the pursuit of truth (they may be good for stoking circulation). Innuendo and repetition rather than conjecture and refutation are often the order of the day. Any genre of writing that aims at conveying truth – reporting pre-eminently among them – needs strong conventions to enable readers to decode and assess what they are reading. These conventions include practices such as setting out the evidence, indicating its sources, and explaining its limitations, as well as structures that provide (for example) for forms of quality control, external assessment or peer review and for sanctions on those who invent or falsify evidence. Reporters would be helped in their proper task if they were supported by regulatory structures and editorial policies that made it less easy to distort and ignore information, to neglect important topics or to highlight trivia. Few British newspapers today could plausibly see themselves as offering a neutral arena for a great game of ideas, or claim that they aim to report what they judge to be the case and to avoid what they judge to be false. Few could pretend that their reporting on medicine, science and biotechnology is aimed at enabling their readers to gain an accurate picture of these areas.

Neither version of the argument for freedom of the press as a means to discovery of truth supports *unrestricted* press freedom. In particular, neither justifies an unrestricted version of press freedom that permits (let alone encourages) a simulacrum of reporting that mixes competent reporting with material that misinforms, disinforms, trivialises or marginalises. A commitment to greater accuracy and to better coverage, would also help rather than harm serious investigative journalism, which is undermined by practices of communication that pay no attention to readers' need to assess what is reported.

On a second argument, freedom of the press is seen as necessary to protect freedom of expression. This is a very important argument for individuals, for whom individual autonomy within the limits of their own obligations, and thereby of respect for others' rights, is of great value and supports religious as well as political freedom. However there is (to the best of my knowledge) no convincing argument for ascribing or assigning unrestricted rights to freedom of expression to institutions, or in particular to media conglomerates and organisations. Companies and other institutions need sufficient freedom of expression to carry out their legitimate purposes. The media need sufficient freedom to carry out their legitimate purposes, and that includes freedom to present unpopular and derided views. They may also, like any other entertainment business, choose to publish material that makes no claim to truth: bridge problems and competitions, short stories and book reviews, commentary and analysis do not deceive because they do not aim at reporting truth. What deceives is the manipulation of purported reporting. The fact that the media need this *wide freedom* to publish does not show that they have a right to unrestricted freedom of expression, let alone a right to use that freedom of expression to abandon reliable reporting.

On a third view, the reason for fostering freedom of the press is to support the public interest in achieving what is often called in US discussions of the First Amendment[22] 'wide-open, robust debate'. This is a powerful argument for freedom of the press as means to ends that are of great importance in any democracy; but it too does not support *unrestricted* freedom of self-expression. In particular, it gives no support to uses of freedom of the press that undermine rather than support wide-open, robust debate. It therefore gives no support to editorial policies or journalistic practices that fail to cover important matters of deep public concern, or that suggest that trivial issues are of high importance. In particular, it gives no support to editorial or journalistic strategies that are cavalier about evidence in medicine, science and biotechnology or credulous about 'stories' and 'scares'. It is this third argument, I believe, that is most fundamental for justification of freedom of the press: and it too is not an argument that supports an unrestricted right to

[22] The First Amendment to the US Constitution reads: 'Congress shall make no law respecting an establishment of religion, or prohibiting the free exercise thereof; or abridging the freedom of speech, or of the press, or the right of the people peaceably to assemble, and to petition the Government for a redress of grievances.' It remains the starting point for countless US discussions of press freedom.

freedom of expression for the press, although it provides additional support for a wide right to freedom of expression for individuals.

These three arguments have long histories. They do not individually or collectively support unrestricted freedom of the press. The only argument that supports (more or less)[23] unrestricted freedom of expression is the appeal to individual autonomy or self-expression. That argument will not establish unrestricted freedom of expression for the media, or for other institutions, unless we can show that institutions are entitled to individual autonomy and rights of self-expression. It is unlikely that this can be done. On the contrary, there are good reasons to think that even natural persons have only limited rights to individual autonomy or self-expression,[24] and that corporate persons – whose overweening exercise of autonomy or self-expression can be so damaging to others and their rights – have even less extensive rights to self-expression.

8.6 PRESS RESPONSIBILITIES AND BIOETHICS

Rights to individual autonomy and self-expression do not offer a satisfactory basis for bioethics; they also cannot offer a satisfactory basis for media ethics. Rather than looking at the rights that media companies and conglomerates ought to enjoy, we may get further by considering their obligations. This reversal of perspective is appropriate for media ethics: the media are evidently in the business of *communication*, not of *self-expression*.

Parts of the media may claim – with some candour – to be in the business of communicating only for entertainment, persuasion or for profit, rather than to inform, hence exempt from any responsibility for accurate reporting. The claim is hard to reconcile with other aspects of their self-presentation and practice. And the concession – if that is what it is intended to be – comes at high price: it undermines the most impressive argument for press freedom

[23] The caveat covers traditional limits on individual freedom of expression, generally held to exclude the right to shout 'Fire!' in a crowded theatre, or hate speech, or libel and the like.

[24] Chapters 3 and 4.

as contributing to a 'wide-open robust debate'. In any case the concession cannot establish that the media are exempt from obligations that apply to all other agents and institutions, including obligations to reject deception and coercion.

Communication is always directed to an audience, so rightly subject to ethical requirements rather than a matter for mere, sheer choice and self-expression. An account of freedom of the press should therefore start from a view of the obligations and rights, among them the freedoms, that are relevant to all communication, and thereby also to all mediated communication. If we start from a conception of principled autonomy, policies and principles that deceive must be rejected. Rejecting deception sets a demanding standard. It views speech and writing not as mere self-expression, but as communication. It focuses not on the individualistic question 'how may I (or we) express myself (ourselves)?' but on the broader ethical and social question 'how ought, and may, I or we communicate?'. Communication differs from mere self-expression, which may be oblivious of its audience; mediated communication differs in aiming at a wide and diverse public audience, who may be greatly influenced by what they read and see. Those who aim to communicate are obliged to attend to the characteristics and capacities of their intended audiences, and to take account of what is needed to address them in ways that eschew deception.

Evidently communication will be inadequate if it is unintelligible to its intended audiences. Unintelligibility is not the most common failing of British newspapers. They often use the tools of communication competently, and sometimes brilliantly. A more typical failing is that they neglect or breach fundamental ethical obligations, and in particular obligations to reject deception. This failure has many forms. At its worst it includes writing and programme-making that purports to report yet fails to provide salient information, publishes misinformation, or even disinformation, disguises advocacy as reporting and omits (ranges of) significant evidence.

Standards of accuracy are sometimes particularly poor in coverage of bioethics, where some topics lack universal interest, so do not receive the scrutiny that reporting on national and international politics, on business or on sport attracts. Yet these are topics where accurate and balanced reporting is important for the public. It is very common for such topics to be cast merely as 'human interest stories', with an implied freedom to set aside issues that go beyond a particular case, however atypical, and however important the wider issues. For example, programmes on medicine, science and biotechnology often juxtapose heartrending images (child with cancer; pristine natural environment and foul effluents) with hidebound official spokespersons. Such 'confrontations' supposedly make good 'stories'; they often do not help viewers to make sense of medical or environmental issues.[25]

Most fundamentally, in my view, there is a habitual failure to provide readers with any means of checking and interpreting what they are reading. Even journalists who report reasonably accurately need not provide readers with the possibility of judging what they report and working out whether it can be trusted. When they fail to do so they communicate in ways that are needlessly deceptive, so flout fundamental obligations. *Communication, unlike mere self-expression, is ethically acceptable only when it aims to be accessible to and assessable by its audiences.*[26]

Communicating in ways that intended audiences can both follow and assess can be demanding. In ordinary face-to-face conversation we assess what we are told by backtracking and asking questions, by cross-checking and testing our understanding and our interlocutors. We build up a picture of what others mean; where relevant we judge the truth of their claims and of the advisability of trusting them. Because written and broadcast communication is almost exclusively one-way, any writer or broadcaster who rejects deception has to supply far more to enable readers to check and cross-check, and to establish a basis for judging whether to place

[25] With thanks to two Edinburgh programme-makers who pointed out to me after I lectured on these topics that I had overlooked how standard this tactic for misleading has become in television programmes on medicine, science and biotechnology, and how misleading it can be.

[26] For more detailed arguments see Onora O'Neill, 'Practices of Toleration' in Judith Lichtenberg, ed., *The Media and Democratic Values*, Cambridge University Press, 1990, 155–85.

or refuse trust. Such conventions and standards are the backbone of practices of one-way communication in which truth matters, *a fortiori* of those that bear on trust.

If the media had *unrestricted* rights of self-expression they would be exempted from the standard disciplines of one-way communication: they would have a unique exemption from the basic obligations of communication in matters where truth matters. Writing and broadcasting would incur no criticism even if it were impossible to assess its reliability or adequacy. It would be entirely acceptable if the possibility of reaching reliable judgements about medicine, science and biotechnology were denied to most of us and confined to a well-informed elite with access to other, more reliable sources of information.[27]

If the claims and the configuration of freedom of the press are based, as I think they should be, mainly on their potential contribution to 'wide-open, robust debate' among the public at large, then this overriding purpose must be met rather than ignored. And that requires standards for reporting, including for reporting on medicine, science and biotechnology, that allow readers both to understand and to assess what they read. Meeting these requirements will not prevent ethically responsible newspapers from publishing fiction and horoscopes, recipes and think pieces: all of these can be presented as what they are, and any lack of truth claims can (and should) be made apparent. It shows simply that anybody, and any newspaper or programme, that presents some of its content as reporting has no unrestricted right to communicate untrustworthily, for example by substituting factoids and half-truths, rumours and rubbish for reporting, knowing that at least some (and probably many) readers cannot hope to disentangle them from reliable reporting. It shows that there is no right to report without covering significant aspects of issues, to misreport

[27] Not everyone is in this happy position. If they were, the lowest standards of reporting could be defended on the bizarre grounds that no deception occurs, because the entire readership knows enough not to be taken in! Defences – or excuses – that the contents of a publication or programme on matters of public policy are 'only entertainment' sometimes approach this Alice in Wonderland level.

others' views, or to 'report' falsehoods. It shows that there can be no warrant for conflating gossip with news, or invention with reporting. Corporate communicators have no rights to exemption from these requirements. (There may be good reasons for rather greater indulgence when individuals mislead: they are less likely to be taken as authoritative, and are often disciplined to a degree by fear of losing others' respect and trust.)

The media, of course, are not the only exponents of inaccurate, slipshod and deceptive communication. Politicians and businesses, campaigning organisations and the public relations 'experts' they all hire are constantly spinning 'stories', hoping that the media will present matters with their spin and nobody else's. When they do this, the standards of their communication are ethically unacceptable. In mitigation it might be said that, even if the most powerful can sometimes 'put across' a lying version of events, many less powerful office holders and institutions act in self-defence. They dare not speak without spin in face of a media culture in which they have come to expect to be misrepresented without redress, so aim to pre-empt misrepresentation by persuading the media to their way of thinking. The media hold much of the remedy to the spin of which they so amply accuse others in their own hands: they alone have the power to render spinning ineffective.[28]

This dire situation matters a great deal for public discussion of medicine, science and biotechnology. These are difficult and complex activities that bear on everyone's life. They are not on a par with gossip about the lives of those who choose to bare their breasts (or other bits) to journalists for their own self-publicising purposes. A commitment to reject deception is therefore likely to make particularly clear demands in communicating about medicine, science and biotechnology. Doing so to adequate standards is likely to require good journalists who can write well and are equipped with enough specialist knowledge to understand the topics on which

[28] Of course there is a problem of collective action here. Perhaps nobody dares to get a grip on the practice of publicising 'stories' with spin, knowing that others will take advantage, and that there are no truces in circulation wars. This, of course, is a classic reason for imposing regulation.

they write and to discriminate among sources, as well as editors and proprietors who insist that reporting be anchored in competence and done to a good standard. It is likely to require an end to practices of handing technical reporting to political correspondents without the necessary competence (the converse handover is less popular).[29] It is likely to require self-restraint in covering individual cases that have a ludicrous or sentimental aspect ('health' reporting is full of this sort of 'story'). It requires either an end to chequebook journalism, or at least full disclosure of who was paid how much by whom for which 'information' in writing a 'story' (this requirement would presumably curtail the practice). It is likely to require a more systematic attempt to indicate within any piece of writing the sorts of evidence for its various elements, and quite explicit indications of any element of conjecture, let alone invention. It would curtail the fun, malice and sheer irresponsibility of writing misleading headlines.

Of course, few of us are brave enough to say that this not merely ought to be done, but that it ought to be enforced. Perhaps we should be braver. Many improvements along these lines can be introduced without trace of censorship. Censorship is a matter of forbidding or requiring publication of certain content: there can be no case for it, or for putting an end to investigative journalism. On the contrary, change is essential if genuine investigative journalism is not to be smothered by irresponsible and inadequate imitations. A free and responsible press is incompatible with controls on content, but it ought and can be subject to procedural requirements. Newspapers that met appropriate requirements could be better than many now are, and the standards of print journalism could at least be brought closer to those of broadcast journalism that complies with the regulations to which it is subject (while also reducing the downward pressure that abysmal newspaper standards can exert on broadcasting standards and on their readership's expectations). It is fully compatible with freedom of the press to insist on ethically acceptable processes of reporting, as is demanded of the

[29] House of Lords, Select Committee on Science and Technology, *Report on Science and Society*, HL 38, 2000, includes evidence on this point; see institutional bibliography.

broadcast media in the USA, the UK and in many other countries (if sometimes poorly respected or enforced). Responsible writing about science, medicine and biotechnology can be entirely free from censorship, while facing at least some of the disciplines of accountability that structure other areas of life, such as the public sector and the professions.

What improvements in process might be demanded? I start cautiously with the following suggestions. Owners, editors, programme-makers and reporters might be required to declare and publish relevant interests and conflicts of interests in a prominent and informative form. For them, as for others, these declarations could include financial and non-financial interests, and perceived interests, including those arising through family and personal relationships. They could be required to declare relations with lobbyists, political parties, companies and campaigning organisations prominently, publicly and intelligibly in articles or programmes to which they pertained. Just as it matters to readers to know whether health advice is given by a doctor, by an enthusiast for home or exotic remedies or by a spiritual healer, so it matters to them to know whether an 'environmental correspondent' is also active on behalf of a biotech company or an environmental campaigning organisation. The media could be required to publish the credentials of reporters writing on technical topics, and to warn and comment if they assign reporters lacking the relevant competence to write on a technical subject. They could be required to declare full financial information about payments made to obtain material relevant to 'stories', alongside each and every published version of the 'story'. Where they had published misinformation or disinformation, publication of corrections of equal length and prominence, perhaps written by third parties, could be required. Penalties for quoting, thereby recirculating, 'stories' shown to be libellous or invented could be used.

Good reporters, good editors and good proprietors often try to keep to these and more exigent standards; but since these practices are not widely followed, and since flouting them is deceptive, there is a strong case for enforcing these standards. The point is simple.

Any public debate on medicine, science and biotechnology – or on other areas of life – is likely to end up undermining trust so long as one powerful interest group is allowed to misrepresent itself as reporting, or even as nobly speaking truth to power, and yet is exempt from any analogue of the disciplines that others – rightly – are required to face. I am reasonably sure that a largish improvement could be achieved without any censorship, since these proposals constrain the process but not the content of reporting. All that is required is living and writing by the standards that many journalists would recognise as important. On the other hand, while some proprietors, editors and programme-makers provide ethically inadequate leadership, or even push poor practices to their limits, and unrestricted press freedom is *de facto* accepted, little will change.

Unless standards of reporting are improved, either from within or from without, we cannot expect any increase in public trust in medicine, science and biotechnology. It will not matter how far trustworthiness is improved in the public sector and in the professions, or whether there is reasonably good evidence that the majority of doctors and scientists are trustworthy, or whether the biotech industry in the UK remains (in many respects) tightly regulated, or whether there is a strong culture of compliance with regulatory demands in the UK, or whether those who abuse trust are disciplined: there will be no return of public trust. No amount of effort to improve trustworthiness, and no amount of investigative journalism that eschews deception, will be able to counteract a media culture in which deception is permitted and routine. If the media 'reports' without any regard for communicative obligations, if it indiscriminately labels office holders, professionals and experts inadequate or untrustworthy whenever a colleague fails, and regardless of evidence, if such failure is represented as common or even as the norm when it is not, there will be no return of public trust.

Trust will be restored only if the public have ways of judging matters *for themselves*. And if the only feasible way for most people to reach those judgements is damaged and undermined by a media culture that does not bother to make its reporting assessable by its readers, it seems unlikely that much will change. The results may not be universal alertness and suspicion: just as the shepherd boy who shouted 'Wolf! Wolf!' once too often lost credibility, so a media that announces scandal and suspicion at every turn is readily discounted. Where everybody is somebody, nobody is anybody; and where everything is scandalous, nothing is really scandalous. The likely results will be continued widespread expressions of public distrust in medicine, science and biotechnology, coupled with seemingly inconsistent placing of trust in many office holders, professionals and experts.

If we are optimistic we may hope that it will be enough if medicine, science and biotechnology are made as trustworthy as is feasible, so that trust is in fact often well placed. However, since nothing can guarantee total trustworthiness, a culture that undermines the basis for discrimination in placing trust is heading for problems. I do not think we should underestimate the burden that being commonly misrepresented as untrustworthy and as mistrusted now places on all office holders, professionals, experts and others, a burden that is not lifted when they are trustworthy and comply with all regulatory, professional and audit regimes. Nor should we underestimate the anxiety that well-fomented and publicised uncertainty about others' trustworthiness creates for all citizens.

If we continue to value individual autonomy, and especially the individual autonomy of those who own, direct and work for the media, more highly than we value trust, we should not be surprised that trust remains elusive. In particular we should not be surprised that those doctors, scientists and biotechnologists, and those who fund and regulate them, are so regularly and so indiscriminately represented as untrustworthy. If we take principled autonomy seriously we must also take obligations to reject deception seriously; and once we take the rejection of deception seriously we will have reason to build – or rebuild – institutions that help us to discriminate between cases. In building those institutions and practices we would foster rather than undermine relations of trust, and would allow individual autonomy its due place, but no more.

Bibliography

BOOKS AND ARTICLES

Adams, John, *Risk*, UCL Press, London, 1995

Adorno, Theodore W., *The Authoritarian Personality*, Harper & Bros, New York, 1950

Alston, Philip, 'International Law and the Human Right to Food', in Philip Alston and K. Tomaševski, eds., *The Right to Food*, Nijhoff, Dordrecht, 1984, 9–68

Andrews, Lori B., 'Informed Consent Statutes and the Decision-Making Process', *Journal of Legal Medicine*, 5, 1984, 163–217

Archard, David, *Children: Rights and Childhood*, Routledge, London, 1993

Armour, A., 'The Citizens' Jury Model of Public Participation: A Critical Evaluation', in O. Renn, T. Webler and P. Wiedemann, eds., *Fairness and Competence in Citizen Participation: Evaluating Models for Environmental Discourse*, Kluwer, Dordrecht, 1995, 175–88

Baier, Annette, 'Trust and Antitrust', *Ethics*, 96, 1986, 231–60

Bainham, Andrew, Sclatel, Shelley Day and Richards, Martin, eds., *What is a Parent? A Socio-Legal Analysis*, Hart Publishing, Oxford, 1999, 121–41

BBC, *Producers' Guidelines: The BBC's Values and Standards*, BBC, London, 2000

Beauchamp, Thomas L. and Childress, James F., *Principles of Bioethics*, Oxford University Press, New York, 1989

Beck, Ulrich, *Risk Society*, Sage, London, 1986

Bentley, Gillian R. and Mascie Taylor, C.G. Nicholas, eds., *Infertility in the Modern World: Present and Future Prospects*, Cambridge University Press, Cambridge, 2000

Bettelheim, Bruno, *A Good Enough Parent: The Guide to Bringing Up Your Child*, Routledge, London, 1995

Blackburn, Simon, *Ruling Passions: A Theory of Practical Reasoning*, Clarendon Press, Oxford, 1998

Buchanan, Allen, Brock, Dan W., Daniels, Norman and Wikler, Daniel, *From Chance to Choice: Genetics and Justice*, Cambridge University Press, Cambridge, 2000

Brownlie, Ian, ed., *Basic Documents on Human Rights*, Clarendon, Oxford, 1981, 21–7

Callahan, Daniel, 'Autonomy: A Moral Good, not a Moral Obsession', *Hastings Center Report*, 14, 1984, 40–2

Callahan, Daniel, 'Can the Moral Commons Survive Autonomy?', *Hastings Center Report*, 26, 1996, 41–2

Cavalieri, Paola and Singer, Peter, *The Great Ape Project: Equality Beyond Humanity*, Fourth Estate, London, 1993

Chadwick, Ruth, ed., *Ethics, Reproduction and Genetic Control*, Croom Helm, London, 1987; revised edn., Routledge, London, 1992

Chadwick, Ruth, Levitt, Mairi and Shickle, Darren, eds., *The Right to Know and the Right Not to Know*, Avebury, Aldershot, 1997

Chadwick, Ruth and Ngwena, Charles 'The Human Genome Project, Predictive Testing and Insurance Contracts: Ethical and Legal Responses', *Res Publica*, 1, 1995, 115–29

Chadwick, Ruth and Thompson, Alison, eds., *Genetic Information Acquisition, Access, and Control*, Kluwer Academic/Plenum Publishers, New York, 1999

Chadwick, Ruth, Darren Shickle, Henk ten Have and Urban Wlesing, eds., *The Ethics of Genetic Screening*, Kluwer, Dordrecht, 1999

Christman, John, 'Constructing the Inner Citadel: Recent Work on the Concept of Autonomy', *Ethics*, 99, 1988, 109–24

Christman, John, ed., *The Inner Citadel: Essays on Individual Autonomy*, Oxford University Press, New York, 1989

Clarke, Adele E., *Disciplining Reproduction: Modernity, American Life Sciences and 'The Problem of Sex'*, University of California Press, Berkeley, 1998

Clarke, Stephen R. I., *The Moral Status of Animals*, Oxford University Press, New York, 1977

Coady, C. A. J., *Testimony*, Oxford University Press, Oxford, 1992

Cook, Rachel, 'Donating Parenthood: Perspectives from Surrogacy and Gamete Donation', in Andrew Bainham, Shelley Day Sclater and Martin Richards, eds., *What is a Parent? A Socio-Legal Analysis*, Hart Publishing, Oxford, 1999, 121–41

Curtice, John and Jowell, Roger, 'Trust in the Political System', in Jowell, Roger *et al.*, eds., *British Social Attitudes: The 14th Report*, Dartmouth, Aldershot, 1997, 89–109

Daniels, Ken, 'The Semen Providers', in Ken Daniels and Erica Haimes, eds., *Donor Insemination: International Social Science Perspectives*, Cambridge University Press, Cambridge, 1998, 76–104

Daniels, Norman, *Just Health Care*, Cambridge University Press, 1985; new edition, *Just Health*, forthcoming

de Beaufort, Inez, 'Letter from a Postmenopausal Mother', in John Harris and Søren Holm, eds., *The Future of Human Reproduction: Ethics, Choice and Regulation*, Clarendon Press, Oxford, 1998, 238–47

Deech, Ruth, 'Cloning and Public Policy', in Justine Burley, ed., *The Genetic Revolution and Human Rights* (The Oxford Amnesty Lectures), Oxford University Press, Oxford, 1999, 95–100

Descartes, René, *Discourse on the Method of Rightly Conducting One's Reason and Seeking the Truth in the Sciences*, *Philosophical Writings of Descartes*, vol. 1, trans John Cottingham, Robert Stoothof and Dugald Murdoch, Cambridge University Press, Cambridge, 1985

Dex, Shirley and Sheppard, Elaine, *Perceptions of Quality in Television Production*, Judge Institute for Management Studies, Cambridge, 2000

Dworkin, Gerald, *The Theory and Practice of Autonomy*, Cambridge University Press, Cambridge, 1988

Dworkin, Gerald, 'The Concept of Autonomy', in John Christman, ed., *The Inner Citadel: Essays on Individual Autonomy*, Oxford University Press, New York, 1989, 54–76

Dworkin, Ronald, *Life's Dominion*, HarperCollins, London, 1983

Dworkin, Ronald, *Freedom's Law*, Oxford University Press, Oxford, 1996

Elster, Jon, 'Sour Grapes – Utilitarianism and the Genesis of Wants', in Amartya Sen and Bernard Williams, eds., *Utilitarianism and Beyond*, Cambridge University Press, Cambridge, 1982, 219–38; reprinted in John Christman, ed., *The Inner Citadel*, 1989, 170–188

Elster, Jon, *Sour Grapes: Studies in the Subversion of Rationality*, Cambridge University Press, Cambridge, 1983

Faden, Ruth and Beauchamp, Thomas, in collaboration with Nancy M. P. King, *A History and Theory of Informed Consent*, Oxford University Press, New York, 1986

Firestone, Shulamith, *The Dialectics of Sex: The Case for Feminist Revolution*, William Morrow & Co, New York, 1970

Fishkin, James, *The Limits of Obligation*, Yale University Press, New Haven, 1982

Fox, Renée, 'More than Bioethics', *Hastings Center Report*, 26, 1996, 5–7

Frankfort, Harry G., 'Freedom of the Will and the Concept of a Person', *Journal of Philosophy*, 68, 1971, 5–20; reprinted in John Christman, ed., *The Inner Citadel*, 1989, 63–90

Frewer, L. J., Howard, C., Heddereley, D. and Shepherd, R., 'What Determines Trust in Information about Food-Related Risks? Underlying Social Constructs', in Ragnar Löfstedt and Lynn Frewer, *Risk and Modern Society* Earthscan, London, 1998, 193–212

Gilland, Tony, 'Precaution, GM Crops and Farmland Birds', in Julian Morris, ed., *Rethinking Risk and the Precautionary Principle*, Butterworth Heinemann, Oxford, 2000, 60–83

Gilligan, Carole, *In A Different Voice: Psychological Theory and Women's Dependence*, Harvard University Press, Cambridge, Mass., 1982; 2nd edition, 1993

Goodin, Robert, 'Democratic Deliberation Within', Philosophy and Public Affairs, 29, 2000, 81–109

Harper, P. S., Lim, C. and Craufurd, D., 'Ten Years of Presymptomatic Testing for Huntington's Disease: The Experience of the UK, Huntington's Disease Prediction Consortium', *Journal of Medical Genetics* 37, 2000, 567–71

Harris, John, *Clones, Genes and Immortality: Ethics and the Genetic Revolution* (revised version of John Harris, *Superman and Wonderwoman*, 1992), Oxford University Press, Oxford, 1998

Harris, John, 'Rights and Reproductive Choice', in John Harris and Søren Holm, eds., *The Future of Human Reproduction: Ethics, Choice and Regulation*, Clarendon Press, Oxford, 1998, 5–37

Harris, John, 'Clones, Genes and Human Rights', in Justine Burley, ed., *The Genetic Revolution and Human Rights* (The Oxford Amnesty Lectures), Oxford University Press, Oxford, 1999, 61–94

Harris, John and Holm, Søren, eds., *The Future of Human Reproduction: Ethics, Choice and Regulation*, Clarendon Press, Oxford, 1998

Harris, Nigel G. E., 'Professional Codes and Kantian Duties', in Ruth Chadwick, ed., *Ethics and the Professions*, Amesbury, Aldershot, 1994, 104–15

Hayward, Tim, 'Anthropocentrism: A Misunderstood Problem', *Environmental Values*, 6, 1997, 49–63

Hill, Thomas E., Jr., 'The Kantian Conception of Autonomy' in Hill, *Dignity and Practical Reason in Kant's Moral Theory*, Cornell University Press, Ithaca, NY, 1992, 76–96

Holton, Richard, 'Deciding to Trust, Coming to Believe', *Australasian Journal of Philosophy*, 72, 1994, 63–76

Hope, R. A. and Fulford, K. W. M., 'Medical Education: Patients, Principles, Practice Skills', in R. Gillon, ed., *Principles of Health Care Ethics*, John Wiley, London, 1993

Humphries, John, *The Great Food Gamble*, Hodder & Stoughton, London, 2001

Husak, Douglas, 'Liberal Neutrality, Autonomy and Drug Prohibition', *Philosophy and Public Affairs*, 29, 2000, 43–80

Jacobs, Lawrence, Marmor, Theodore and Oberlander, Jonathan, 'The Oregon Health Plan and the Political Paradox of Rationing: What Advocates and Critics Have Claimed and What Oregon Did', *Journal of Health Politics, Policy and Law*, 24(1), 1999, 161–80

Johnson, Samuel, *The Rambler* (1750), ed. W. J. Bate and Albrecht B. Strauss, in *The Yale Edition of the Works of Samuel Johnson*, John H. Middendorf, ed., Yale University Press, New Haven, CT, 55

Jones, Karen, 'Trust as an Affective Attitude', *Ethics*, 107, 1996, 4–25

Kant, Immanuel, *Critique of Pure Reason* (1781), trans. Paul Guyer and Allen W. Wood, Cambridge University Press, Cambridge, 1998

Kant, Immanuel, *What is Enlightenment?* (1784), trans. Mary J. Gregor, in Kant, *Practical Philosophy*, Cambridge University Press, Cambridge, 1996, 8:35–42

Kant, Immanuel, *Groundwork of the Metaphysic of Morals* (1785), trans. Mary J. Gregor, in Kant, *Practical Philosophy*, Cambridge University Press, Cambridge, 1996, 4:387–463

Kant, Immanuel, *What Does it Mean to Orient Oneself in Thinking?* (1786), trans. Allen W. Wood and George di Giovanni, in Kant, *Religion and Rational Theology*, eds. Allen W. Wood and George di Giovanni, Cambridge University Press, Cambridge, 1996, 8:133–46

Kant, Immanuel, *Critique of Practical Reason* (1788), trans. Mary J. Gregor, in Kant, *Practical Philosophy*, Cambridge University Press, Cambridge, 1996, 5:19–163

Kant, Immanuel, *The Conflict of the Faculties* (1798), trans. Mary J. Gregor and Robert Anchor, in Kant, *Religion and Rational Theology*, eds. Allen

W. Wood and George di Giovanni, Cambridge University Press, Cambridge, 1996

Kimmelman, J., 'The Promise and Perils of Criminal DNA Data Banking', *Nature Biotechnology*, 18, 2000, 695–6

Kitcher, Philip, *Lives to Come: The Genetic Revolution and Human Possibilities*, Simon & Schuster, New York, 1996

Kohlberg, Lawrence, *The Philosophy of Moral Development*, Harper & Row, San Francisco, 1981

Lipton, Peter, 'The Epistemology of Testimony', *Studies in the History and Philosophy of Science*, 29, 1998, 1–31

Marteau, Theresa and Richards, M. P. M., eds., *The Troubled Helix – Psychosocial Implications of the New Human Genetics*, Cambridge University Press, Cambridge, 1996

Martin, P. and Kaye, J., 'The Use of Biological Sample Collections and Personal Medical Information in Human Genetic Research', *New Genetics and Society*, 19, 2000, 165–92

Milgram, Stanley, *Obedience to Authority: An Experimental View*, Tavistock Publications, London, 1974

Mill, John Stuart, *On Liberty* (1863), in Mary Warnock, ed., *Utilitarianism, On Liberty and other Essays*, Fontana, London, 1962

Morris, Julian, 'Defining the Precautionary Principle', in Julian Morris, ed., *Rethinking Risk and the Precautionary Principle*, Butterworth Heinemann, Oxford, 2000, 1–21

Morris, Julian, ed., *Rethinking Risk and the Precautionary Principle*, Butterworth Heinemann, Oxford, 2000

Murdoch, Iris, *The Sovereignty of the Good*, Routledge & Kegan Paul, London, 1970

Murray, Thomas H., *The Worth of a Child*, University of California Press, Berkeley, Cal., 1996

Murray, Thomas H., 'Genetic Exceptionalism and Future Diaries: Is Genetic Information Different from Other Medical Information?', in Mark A. Rothstein, ed., *Genetic Secrets: Protecting Privacy and Confidentiality*, Yale University Press, New Haven, 1997

North, Richard D., 'Science and the Campaigners', *Economic Affairs*, Institute of Economic Affairs, London, 2000, 27–34

Nussbaum, Martha, *Women and Human Development: The Capabilities Approach*, Cambridge: Cambridge University Press, Cambridge, 2000

Nye, Joseph S., Zelikow, Philip D. and King, David C., eds., *Why People Don't Trust Government*, Harvard University Press, Cambridge, Mass, 1997

O'Neill, Onora, 'Children's Rights and Children's Lives', *Ethics*, 98, 1988, 445–63

O'Neill, Onora, 'Practices of Toleration', in Judith Lichtenberg, ed., *The Media and Democratic Values*, Cambridge University Press, Cambridge, 1990, 155–85

O'Neill, Onora, 'Medical and Scientific Uses of Human Tissues', *Journal of Medical Ethics*, 22, 1996a, 5–7

O'Neill, Onora, *Towards Justice and Virtue: A Constructive Account of Practical Reasoning*, Cambridge University Press, Cambridge, 1996b

O'Neill, Onora, 'Environmental Values, Anthropocentrism and Speciesism', *Environmental Values*, 6, 1997a, 127–42

O'Neill, Onora, 'Genetics, Insurance and Discrimination', *Manchester Statistical Society*, 1997b, 1–14

O'Neill, Onora, 'Insurance and Genetics: The Current State of Play', *The Modern Law Review*, 61, 1998a, 716–23

O'Neill, Onora, 'Necessary Anthropocentrism and Contingent Speciesism', Symposium on 'Kant on Duties Regarding Non-Rational Nature', *Proceedings of the Aristotelian Society*, supplementary vol., 1998b, 211–28

O'Neill, Onora, 'The 'Good Enough' Parent in the Age of the New Reproductive Technologies', in Hille Haker and Deryck Beyleveld, eds., *The Ethics of Genetics in Human Procreation*, Athenaeum Press, Gateshead, 2000, 33–48

O'Neill, Onora, 'Informed Consent and Genetic Information', *Studies in History and Philosophy of Biological and Biomedical Sciences*, 32, 2001a, 689–704

O'Neill, Onora, 'Practical Principles and Practical Judgement', *Hastings Center Report*, 31, 2001b, 15–23

O'Neill, Onora, 'Autonomy and the Fact of Reason in the *Kritik der praktischen Vernunft*, 30–41', in Otfried Höffe, ed., *Immanuel Kant, Kritik der praktischen Vernunft*, Akademie Verlag, Berlin, forthcoming

O'Neill, Onora, 'Kant's Conception of Public Reason', *Proceedings of the IX Kant Kongress*, de Gruyter, Berlin, forthcoming

O'Neill, Onora and Ruddick, William, eds., *Having Children: Philosophical and Legal Reflections on Parenthood*, Oxford University Press, New York, 1979

Piaget, Jean, *The Moral Judgement of the Child*, trans. Marjorie Gabain, Penguin, Harmondsworth, 1977

Pogge, Thomas W., 'Relational Conceptions of Justice: Responsibilities for Health Outcomes', in Anand Sudhir, Peter Fabienne and Amartya Sen, eds., *Health, Ethics, and Equity*, Clarendon Press, Oxford, forthcoming

Power, Michael, *The Audit Explosion*, Demos, London, 1994

Power, Michael, *The Audit Society: Rituals of Verification*, Oxford University Press, Oxford, 1997

Quine, W. V. O., 'Reference and Modality', in Quine, *From a Logical Point of View: 9 Logico-Philosophical Essays*, Harper Torchbooks, New York, 1953, 139–59

Quine, W. V. O., 'Two Dogmas of Empiricism', in Quine, *From a Logical Point of View: 9 Logico-Philosophical Essays*, 2nd edn., Harper & Row, New York, 1963, 20–46

Raab, Charles D., 'Electronic Confidence: Trust, Information and Public Administration', in I. Th. M. Snellen and W. B. H. J. van de Donk, eds., *Public Administration in an Information Age*, IOS Press, Amsterdam, 1998, 113–35

Rawls, J., *A Theory of Justice*, Harvard University Press, Cambridge, Mass., 1971

Reich, W. T., 'The Word 'Bioethics': Its Birth and the Legacies of Those Who Shaped It', *Kennedy Institute of Ethics Journal* 4, 1994, 319–35

Rhodes, Rosamond and Strain, James J., 'Trust and Transforming Healthcare Institutions', *Cambridge Quarterly of Healthcare Ethics*, 9, 2000, 205–17

Robertson, John A., 'Embryos, Families and Procreative Liberty: The Legal Structures of the New Reproduction', *Southern California Law Review*, 59, 1986, 939–1041

Robertson, John A., *Children of Choice: Freedom and the New Reproductive Technologies*, Princeton University Press, Princeton, 1994

Rothman, David J., *Strangers at the Bedside: A History of How Law and Ethics Transformed Medical Decision-Making*, Basic Books, New York, 1991

Rothstein, Mark A., *Genetic Secrets: Protecting Privacy and Confidentiality in the Genetic Era*, Yale University Press, New Haven, 1997

Ruddick, William, 'Parents and Life Prospects', in Onora O'Neill and William Ruddick, eds., *Having Children: Philosophical and Legal Reflections on Parenthood*, Oxford University Press, New York, 1979, 124–37

Schneewind, J. B., *The Invention of Autonomy: A History of Modern Moral Philosophy*, Cambridge University Press, Cambridge, 1998

Semetko, Holli A., 'Great Britain: The End of News at Ten and the Changing News Environment', in Richard Gunther and Anthony Mughan, eds., *Democracy and the Media: A Comparative Perspective*, Cambridge University Press, and Cambridge, 2000, 343–74

Shapin, Steven, *A Social History of Truth*, Chicago University Press, Chicago, 1994

Silver, Lee M., *Remaking Eden: Cloning, Genetic Engineering and the Future of Humankind?*, Weidenfeld & Nicolson, London, 1998

Singer, Peter, 'Famine, Affluence and Morality', *Philosophy and Public Affairs*, 1, 1972, 229–43

Singer, Peter, *Animal Liberation*, Cape, London, 1976

Singer, Peter, *The Expanding Circle: Ethics and Sociobiology*, Clarendon Press, Oxford, 1981

Skorupski, John, *John Stuart Mill*, Routledge, London, 1989

Slovic, Paul, 'Perceived Risk, Trust and Democracy', in Ragnar Löfstedt and Lynn Frewer, eds., *Risk and Modern Society*, Earthscan, London, 1998, 181–92

Starr, Douglas, *Blood: An Epic History of Medicine and Commerce*, Knopf, New York, 1998

Stewart, John, Kendall, Elizabeth and Coote, Anna, *Citizens' Juries*, IPPR, London, 1994

Strasbourg, Martin, Wiener, Joshua and Baker, Robert with Fein, I. Alan, eds., *Rationing America's Medical Care: The Oregon Plan and Beyond*, Brookings Institution, Washington, DC, 1992

Strathern, Marilyn, 'From Improvement to Enhancement: An Anthropological Comment on the Audit Culture', *The Cambridge Review*, 118, 1997, 117–26

Sztompka, Piotr, *Trust: A Sociological Theory*, Cambridge University Press, Cambridge, 1999

Thompson, John, *Political Scandal: Power and Visibility in the Media Age*, Polity, Cambridge, 2000

Titmuss, Richard M., reissued by Anne Oakley and John Ashton, eds. *The Gift Relationship: From Human Blood to Social Policy* (1970), New York: New Press, 1997

Tröhler, Ulrich and Reiter-Theil, Stella, in cooperation with Eckhard Herych, *Ethics Codes in Medicine: Foundations and Achievements of Codification Since 1947*, Ashgate, Aldershot, 1998

Universal Declaration of Human Rights (1948), reprinted in Ian Brownlie, ed., *Basic Human Rights Documents*, Clarendon, Oxford, 1981, 21–7

Veatch, Robert M., 'Autonomy's Temporary Triumph', *Hastings Center Report*, 14, 1984, 38–40

Veatch, Robert, M., 'Abandoning Informed Consent', *Hastings Center Report*, 25, 1995, 5–12

Walden, George, *The New Elites: Making a Career in the Masses*, Allen Lane, The Penguin Press, London, 2000

Warren, Samuel D. and Brandeis, Louis D, 'The Right to Privacy', *Harvard Law Review*, IV, 1890, 194–219

Weijer, C. and Emanuel, E. J., 'Protecting Communities in Biomedical Research', *Science*, 289, 2000, 1142–4

Weil, Simone, *The Need for Roots: A Prelude to a Declaration of Duties toward Mankind* (1949), trans. A. F. Wills, Routledge & Kegan Paul, London, 1952

Wildavsky, Aaron, *Searching for Safety*, Transitions, Oxford University Press, Oxford, 1988a

Wildavsky, Aaron, 'If Claims of Harm from Technology are False, mostly False or Unproven What Does That Tell us about Science?', chapter 10 in Peter Berger *et al.*, eds., *Health, Lifestyle and Environment*, Social Affairs Unit, London, 1988b

Wildavsky, Aaron, 'Trial and Error versus Trial without Error', in Julian Morris, ed., *Rethinking Risk and the Precautionary Principle*, Butterworth Heinemann, Oxford, 2000, 22–45

Williams, Bernard, *Ethics and the Limits of Philosophy*, Fontana, London, 1985

Wolpe, Paul Root, 'The Triumph of Autonomy in American Bioethics: A Sociological View', in Raymond DeVries and Janardan Subedi, eds., *Bioethics and Society: Constructing the Ethical Enterprise*, Englewood Cliffs, NJ: Prentice-Hall, 1998, 38–59

Wood, Allen W., *Kant's Ethical Thought*, Cambridge University Press, Cambridge, 1999

Institutional bibliography

The publisher has used its best endeavour to ensure that the URLs for external websites referred to in this book are correct and active at the time of going to press. However, the publisher has no responsibility for the websites and can make no guarantee that a site will remain live or that the content is or will remain appropriate.

Advisory Committee on Genetic Testing, *Code of Practice and Guidance on Human Genetic Testing Services Supplied Direct to the Public*
http://www.doh.gov.uk/pub/docs/doh/lodrep.pdf, 1997

Advisory Committee on Genetic Testing, *Genetic Testing for Late Onset Disorders*
http://www.doh.gov.uk/pub/docs/doh/lodrep.pdf, 1998

Agriculture and Environment Biotechnology Commission
http://www.aebc.gov.uk/

Association of British Insurers (ABI), *Genetic testing: ABI Code of Practice*
http://www.abi.org.uk/, 1997

Convention for the Protection of Human Rights and of the Dignity of the Human Being with regard to the Application of Biology and Medicine, Council of Europe, DIR/JUR (96) 14, 1998

deCODE Genetics
http://www.decode.com/resources/ihd/

Department of Health, *Genetic Paternity Testing Services – Code of* Practice
http://www.doh.gov.uk/genetics/paternity.htm, 2000

Department of Health, *Governance Arrangements for NHS Research Ethics Committees*
http://www.doh.gov.uk/research, July 2001

DNAnow.com (Paternity testing)
http://www.dnanow.com

Food Standards Agency
http://www.foodstandards.gov.uk/index.htm

Genetic Interest Group, *Confidentiality Guidelines*
http://www.gig.org.uk/docs/gig_confidentiality.pdf, 19

Genetics and Insurance Committee (GAIC)
http://www.doh.gov.uk/genetics/gaic.htm

Hastings Center
http://www.thehastingscenter.org/main.htm

Home office, *Fingerprints, Footprints and DNA Samples*, London
http://www.homeoffice.gov.uk/ppd/finger.htm, 1999

House of Commons, Select Committee on Science and Technology, *Human Genetics: The Science and its Consequences*, Third Report, HC 231, 1995

House of Commons, Select Committee on Science and Technology, *Report on Genetics and Insurance*, HC 174, 2001
http://www.publications.parliament.uk/pa/cm200001/cmselect/cmsctech/174/17402.htm

House of Lords, Select Committee on Science and Technology, *Report on Science and Society*, HL 38
http://www.parliament.the-stationery-office.co.uk/pa/ld199900/ldselect/ldsctech/38/3801.htm, 2000

House of Lords, Select Committee on Science and Technology, IIa, *Report on Human Genetic Databases: Challenges and Opportunities*, HL57, 2001; written evidence, HL 115, 2000
http://www.parliament.the-stationery-office.co.uk/pa/ld/ldsctech.htm

Human Genetics Advisory Commission, *The Implications of Genetic Testing for Insurance*, Office of Science and Technology, London
http://www.dti.gov.uk/hgac/, 1997

Human Genetics Advisory Commission, *The Implications of Genetic Testing For Employment*, Office of Science and Technology, London
http://www.dti.gov.uk/hgac/, 1999

Human Genetics Commission, *The Use of Genetic Information in Insurance: Interim Recommendations*
http://www.hgc.gov.uk/business_publications_statement_01may.htm, 2001

The Kennedy Report, *Learning from Bristol: The Report of the Public Inquiry into Children's Heart Surgery at the Bristol Royal Infirmary 1984–1995*, CM 5207
http://www.bristol-inquiry.org.uk/, 2001

MORI and The Human Genetics Commission, *Public Attitudes to Human Genetic Information*
http://www.hgc.gov.uk/business_publications_morigeneticattitudes.pdf, 2001

MORI, summaries of numerous recent polls
http://www.mori.com/polls/index_pl.shtml

Nolan Committee, First Report of the Committee on Standards in Public Life, HMSO, London
http://www.public-standards.gov.uk/default.htm, 1995

Nuffield Council on Bioethics, *Genetic Screening: Ethical Issues*
http://www.nuffield.org.uk/bioethics/index.html, 1993

Nuffield Council on Bioethics, *Human Tissue: Ethical and Legal Issues*
http://www.nuffield.org.uk/bioethics/index.html, 1995

Nuffield Council on Bioethics, *Mental Disorders and Genetics: The Ethical Context*
http://www.nuffield.org.uk/bioethics/index.html, 1998

Nuffield Council on Bioethics, *Genetically Modified Crops: The Ethical and Social Issues*
http://www.nuffield.org.uk/bioethics/index.html, 1999

Office of Science and Technology, *Biotechnology Framework Review*
http://www.dti.gov.uk/ost/rmay99/Bioreprt.pdf, 1999

Office of Science and Technology, Scientific Advice in Policy Making
http://www.dti.gov.uk/ost/aboutost/guidelines.htm, 2000

Parliamentary Office of Science and Technology (POST) *The 'Great GM Food Debate': A Survey of Media Coverage in the First Half of 1999*, Report 138
www.parliament.uk/post/home.htm, May 2000

Report of the Citizens' Jury on Genetic Testing for Common Disorders, Welsh Institute for Health and Social Care, University of Glamorgan, Pontypridd
http://www.medinfo.cam.ac.uk/phgu/info_database/Testing_etc/citizens'_jury.asp, 1998

The Phillips Report, *Report of the Inquiry into BSE and variant CJD in the United Kingdom*
http://www.bse.org.uk/, 2000

The Redfern Report, *Report of the Royal Liverpool Children's Inquiry* ('The Redfern Report on events at Alder Hey Hospital') MRC
http://www.rlcinquiry.org.uk/, 2001

Royal College of Pathologists, *Interior Guidelines for use of 'Surplus' Tissue*
http://www.repath.org/activities/publications/transitional/html, June 2001

UK Forensic Database
http://www.forensic.gov.uk/forensic/entry.htm

World Medical Association, Declaration of Helsinki, as revised Edinburgh 2000
http://www.wits.ac.za/bioethics/helsinki.htm

Index

Printed in the United States
By Bookmasters

Printed in the United States
By Bookmasters